Dedication

No one has ever sacrificed more for the people of the United States than the men and women of our military who serve in combat.

It is my goal to have a small part of their history live on in these pages as a testament to these courageous people.

This book is dedicated to tankers of all wars and all branches of the Military. There is something to be said about these original iron men. Some were almost as hard as the metal their tanks were made of. The crews of these behemoths formed an extremely tight fitting group when combined with the tank, became one of the most formidable pieces of fighting equipment to ever enter the battle field.

They didn't have to walk or carry their equipment but once in battle became the center of attention. They became magnets for all sorts of incoming fire. The enemy wanted the big stuff and would get to the smaller stuff later.

I am very proud to have fought beside, know and call friend some of the people who wrote chapters in this book. They are some of the most dedicated and hard-fighting men this country has to offer.

Those who read this book venture into the deepest thoughts, anguish and memories of these brave men whose pain and suffering lasted long after the battle field.

Acknowledgements

I want to thank all the great and courageous people who sent me their stories to put into this book.

I want to thank Sergeant John Wear who served in 3rd Tank Battalion, 3rd Marine Division in Vietnam in 1968-1969. John is also the President of the USMC Vietnam Tankers Association and Editor of the magazine "Sponson Box". John has helped me immensely.

I would also like to thank my daughter, Tina Simmons, who did all the hard work on my first book, Tracks: Memoirs of a Vietnam Veteran.

Sgt. Gary Mefford and his wife Marge spent many hours editing this book and I greatly appreciate his help. Gary and I were friends in Vietnam and have continued our friendship for many years.

Introduction

My name is Sergeant Clyde Hoch. I served in the Marine Corps. from 1965 to 1969 in "C" Company, 1st Tank Battalion, 1st Marine Division. I arrived in Vietnam in the middle of Tet of '68 and had the pleasure of being there for Tet of '69. This was the time of the heaviest fighting of the whole war. I was a Tank Commander and Section Leader and eventually a Platoon Sergeant.

Fame and fortune are not my goal. I wrote this book so future generations would know what it was like to go to war and get to know the people who fought for this country. I feel future wars will be fought by computer operators working out of the United States. My goal is to leave a record of first-hand stories about what it was like to go to combat.

I have the greatest respect for all the people whose stories are included in this book. It was my intention to have stories of Army and USMC tankers. I have contact with mostly Marines, so I have mostly Marine stories.

I also wrote "Tracks: Memoirs of a Vietnam Veteran". It is a book about my military experience.

Table of Contents

American Tanks of WWI
Written by Clyde Hoch

When tanks were first created, the people working on the top secret British projects wanted to know what they were making. They said, "they look like water tanks." The name tank stuck. One of the original names for tanks was land battleship.

The U. S. did not have a tank in production at the time of WWI. There were several prototypes but no production model. Some of the early prototypes were gas-electric and some steam.

The U. S. licensed the French Renault FT- 17 and started to manufacture them, but the war was over by the time they arrived in Europe.

The Ft-17 was a small light tank weighing in at six tons. It had a four cylinder gas engine that put out 42 horse power. Its top speed was five miles per hour. It had a range of 30 miles and a crew of three. The tank usually carried a 37 millimeter main gun. Although small, it was highly successful and thousands were made.

A Renault Ft-17 used by Americans

The U. S. also used British Mark V tanks. The Mark V was developed from the many models of Mark tanks with great improvements. The early Mark tanks had four drivers, the Mark V had one.

The Mark V tanks were made of two types. One was called a Male. It had 57 millimeter main guns on each side of the tank. The ones called Females had machine guns on each side. 400 Mark V tanks were manufactured, 200 male and 200 female.

The Mark V used a six cylinder gas engine, which put out 150 horse power. It had a range of 45 miles and a top speed of five miles per hour. Originally it was meant to carry troops, but because of fumes in the very poorly ventilated early tanks, the troops were unfit for combat once they reached their destination.

Mechanical issues hampered early tanks immensely. Many broke down on route to the battle field. Because of the gasoline there was always a danger of fires, even if the tank was not hit by enemy fire.

British Mark V tank used by Americans in WWI
Photo courtesy of World War I Vets

Sergeant George Noble Irwin

Written by George's son Don Irwin

Photo by WWI vets

My dad grew up on a farm in Albert, Indiana and joined the Army when he was 23 years old in 1917. He finished Army basic training at Camp Taylor, Kentucky on 26 Feb 1918 and volunteered for the new Heavy Tank Corps. He reported to Camp Mead, Maryland on 23 March for tank training. But things changed rapidly and the Army decided to move tank training to Camp Colt, Pennsylvania to train in the French Renault FT-17 tank which was commanded by Captain Dwight Eisenhower.

However, having volunteered for the British Mark V tank, he and his unit, the 65th Engineers (later the 301st Tank Corps) sailed to Brest, France from Camp Merritt, Hoboken, New Jersey on 28 April 1918 and then to

Warham, England for heavy tank training with the Brits. The 301st new commander was Major Ralph I. Sasse and Company C's commander was Captain Ralph de P. Clarke. Dad seemed to alternate back and forth between Warham and Bovington Field, I assume learning to employ the new Mark V Tank with his company and crew of eight.

The 301st Tank Corps shipped out to Le Havre, France on 24 Aug 1918. They sailed with their 46 or 47 (depending on which account you read) British Mark Vs and Mark V*s (later pronounced Mark V Star). The Mark V had a crew of eight and could have had either two six pound cannons (male) or six Hotchkiss machine guns (female). The Mark V* was a Mark V cut in half, with a six foot mid section added to give it a better chance of crossing the 3.5 meter trenches in the Hindenberg Line. It was also thought that it could carry 24 troops, but later it was discovered that due to poisonous air inside the tank, the troops were in no condition to fight when they arrived at the front, so it carried supplies, if anything, after that. Their unit was then attached to the British 4th Tank Brigade for what was to be a very short stint on the Western Front.

During the next two months, the 301st had four engagements against the Germans, with varying degrees of success. The first against the vaunted Hindenberg Line near Le Catalet and St Quentin on 29 Sept. Each Company of the 301st (AB&C) was assigned to lead one of the 27th Division's regiments (107th, 105th & 108th). The simple scheme was to precede the infantry by about a hundred yards to clear wire obstacles and destroy the German machine gun nests, and artillery emplacements. Dad's company (C), with 15 tanks, was attached to the 107th infantry on the left, but due to the regimental commander's refusal to leave on the new H hour (an hour earlier than the old H hour), and training between the Doughboys and tanks, the tanks were 1,000 yards in front of the regiment.

They encountered withering fire from German guns, and heavy smoke and fog. The visibility was so bad that the tank commanders had to dismount, and walk in front of the tanks in the dark through mud, wire entanglements, abandoned trenches and shell holes, to guide their tanks. Both the 310st and 27th received casualties. Sixteen tanks received direct hits, and only five of the original 40 tanks reached the Hindenburg Line, and none could get through those wide trenches. It's estimated that the 107th lost 1000 men. The 301st lost 3 officers and 17 enlisted men. Fifteen officers, and 70 enlisted men were wounded, and seven enlisted men were missing. An ominous beginning.

The second engagement was on 8 Oct 1918, supporting II American Corps, near Brancourt. Due to heavy losses on 29 Sep, the 301st had only 23 tanks with no reserve. Visibility was perfect, with a light breeze taking the battle smoke over the Germans. They were bombed from the air, and received heavy shelling from the Germans, but eleven tanks reached their objective, four received direct hits and were destroyed: five had mechanical trouble: and three didn't make it past the jump off point. Relatively speaking, this battle was a success, with the objective being met by 1415 that afternoon.

The third was fought on 17 Oct 1918 from Le Haie Meneresse to Mazingbien. The tank commanders were able to view the battle field prior to engagement, and made a note of the best routes and river crossings. Twenty five tanks were available. In spite of excellent planning, on the day of the battle the visibility was poor, a light breeze blowing smoke over the Doughboys, and communication was even poorer. By the time the visibility cleared, most tanks were lost and out of gas. Of the 25 tanks that started, eight had mechanical trouble, two tanks ditched in the Selle River, a couple had direct hits, and one caught fire. Only one tank reached the objective. Now there were only twelve tanks serviceable.

The last battle was on 23 Oct 1918. The 301st was allotted to the IX British Corps. H hour was 0120 taking advantage of the full moon. They divided their tanks into three companies of four tanks each. Early on, the tanks could see the routes, and gunners could identify their targets. Later, however, the Germans gassed them, and the use of the gas masks greatly limited visibility. There were no losses this time, and all objectives were met.

The Armistice was signed 11 November 1918. Dad sailed home on 28 February 1919 aboard the P.W. Wilson from Marseille, was discharged on 9 April 1919 and went back to work his farm near Albert, Indiana.

Life in the Mark V was very unpleasant, the air contaminated from poorly ventilated gasses of carbon monoxide and cordite fumes. Loud beyond belief, with temperatures reaching 120 degrees. The crews wore helmets, and masks of chain mail to protect them from pieces of metal and rivets knocked loose from shells hitting the external armor. The tank, to be sure, had a fitful start as a weapon system, and like the rest of the U. S. Army was rapidly dismantled after the war. The tank Corps future was doubtful.

Dad never spoke a word to me about any of this. The only thing I got was a diary, such as it was, with dates and places. The only emotion shown in his diary was written onboard the PW Wilson on the way home on the 28th of February 1919. He wrote, "Just eleven months ago at 10:00 AM, sailed from NY bound for the fight, Happy?"

British Mark V tank used by Americans

Tanks of WWII
Written by Clyde Hoch

The M-4 Sherman tank was the primary tank used by the U. S. during WWII. England also used many M-4 Sherman tanks under the Lend Lease act. The British named their American made tanks after American Civil War generals, hence the name Sherman.

Because of the ease of production over 50,000 were made. It was effective because of sheer number. It was produced from 1942 to 1955.

The M-4 Sherman was a medium tank. It was highly effective in Europe and the islands in the Pacific against German and Japanese light to medium tanks. They had to gang up on the heavy German tanks.

The M-4 Sherman weighed in at about 30 tons. It carried a 75 millimeter main gun on a fully traversable turret. It carried 90 rounds for the main gun. It had a gyrostabilizer so the main gun could be fired with some accuracy on the move. It carried a 50 caliber machine gun and two 30.06 caliber machine guns.

It had a range of 120 miles. It was powered by a Continental air cooled radial V-8 gasoline engine that put out about 400 horse power, at a speed up to 30 miles per hour. The engines were manufactured by different manufacturers like Ford, Chrysler and Continental. It had a crew of five, tank commander, gunner, a driver and a co driver and a loader.

The M-4 Sherman variations were used up to and including the Korean War.

The M-4's were also fitted with flame throwers, (M4A3R3) or Zippos as they were commonly called, that were mostly used in the Pacific Islands.

M-4 Sherman of WWII

Charlie Sherman
Written by Charlie Sherman

Marine Corps Tankers at Jacques Farms Tank School

Editor's Note: Charlie Sherman had originally hand-written this story with the intention of it appearing in the Marine Corps Tankers Association newsletter. As inevitable events progressed, a portion of it was actually published but then a change in editorial staff and policy dictated that Charlie's story ended up in "storage" until now. We are proud to include Charlie's personal account in this book and we thank him for his dedication and service to our wonderful country.

A Marine's Story
Written by Charlie Sherman

I recall one day our Senior Drill Instructor came into the Quonset hut to hold an inspection. The first man on the right had not made his rack properly and the DI lit into him something terrible. All of a sudden the recruit hauls back and hits the DI with a round house punch and knocks the DI out the door. Boy! Was there hell to pay for all of us! But not the recruit who hit the DI. The MPs came and took him away and we never saw him again.

When we were at the rifle range, I shot a 343 on "pre-record day". I grew up in the country shooting a rifle since I was a young boy. We had one recruit from out east somewhere who was afraid of his rifle. All he ever got were Maggie's drawers (when you shoot at a target and completely miss it). My DI asked me to go ahead and qualify the next day and then shoot into the recruit's target to allow him to qualify. The reason for this was if our platoon had 100% qualifiers on the range, the DI would get a week off. Well, I fired my target and got a 394. I then fired his target and I beat my own score by one point for a 395! I sure hope that kid did not wear his Expert Rifleman Badge!

Jacques Farm
Written by Charlie Sherman

After boot camp, we went to Jacques Farm for tank school. Most of us recall how nicely we got our shoes to shine in Boot Camp. Well, when the truck stopped at Jacques Farm, it stopped in the middle of a big mud puddle. We jumped off of the truck and landed right in it. To make matters worse, they threw our sea bags down to us and most of the sea bags landed in the mud puddle as well. What a mess!

We went to our assigned tent, unpacked our gear and changed into our utility uniforms. The mess hall was an old cow barn and we ate in our mess kits. My buddy Joe Weitzel and I had not cleaned our mess kits very well and they were pretty dirty. We walked in to the mess hall and surveyed the place. The line that we were to walk down to receive our food (if you can call it that) was the center line where the cow manure would be collected and pushed out the barn. As we walked down the line, the Marines on mess duty simply threw the food toward our mess kits, much of the time missing the mess kit and spattering it in the trough that was for the cow manure. When Joe and I got out of the mess hall, there were two garbage barrels standing there. We both simply dumped our chow, mess kits and all, into the barrels. We both had $40 in our pockets so we went to the PX and ate there until we ran out of money.

Joe and I were at the head and then the wash stand doing our laundry when a run away tank came crashing down onto the camp. I had never heard such screaming as that tank engine made that day. I always wondered how many Marines got hurt that day.

M-4 Sherman tanks at Jacques Farms Tank School - later became Camp Elliot

Going to War
Written by Charlie Sherman

On the way from San Francisco to Honolulu, we were on the ship "Acaquoine". The captain of the ship said he had made the run between the two ports 52 times. He bragged that he had never gotten sea sick. Well, let me tell you, the seas were so rough that everyone including the captain got sick on this trip. There were men throwing up everywhere. Some men were so weak that they could not hold onto anything. As the ship pitched around, these men were thrown against the bulk head and were really injured. The troop compartment had bunks five or six high. There were men on the top bunks who were so sick and weak that they would simply stick their heads over the sides of their bunk and throw up on the bunk below them. The ship's heads were a mess. There was slime everywhere. This lasted two or three days until the seas calmed down a little. Me? I got a little sick one day but as soon as I went top side and got some fresh air, I recovered fairly quickly.

Iwo Jima
Written by Charlie Sherman

10 February 1945 to 16 March 1946

I was on an APA troop ship on the way to Iwo but at the time we did not know where we were headed. On the ship, they gave me a .30 cal machine gun and told me to learn how to take it apart and put it back together blind folded. I did not know why they did this. I learned it in a day or two so they had me teaching other Marines how to do it. When we got to Guam and Siapan, they gave me a Higgins boat and three men to go ashore for the mail. Before we left, one of the men on board (I can't remember if he was a Marine or a sailor) asked me to do him a favor. He explained that before the war he had been a dealer in Las Vegas. He played cards all the time and made a lot of money. He asked me to take his money, buy money orders, make them out to his wife and mail them to her. I did this task almost every day we were in port. He told me it would be just his luck that he would be killed and some other man would enjoy all of the money. I never knew if he made it or not. His name was Tomas T. Turner.

On the day that we landed on Iwo Jima, I found out why they had me working with the .30 cal machine gun. The first boat headed for Red Beach was being loaded with Marines and they told me to also load up on it. I was to provide the .30 cal machine gun cover for the landing. We landed at 0930, unloaded the Marines on the beach and headed back for a second load.

All hell had broken out on the beach by the time we got back to the ship. It was not until 1600 or 1700 that we were able to unload our second boat load of Marines.

Then the waves got so bad that we could not get off the shore. In the mean time, the Japanese were shooting at us with pretty accurate fire. Our Higgins boat got some pretty bad damage and we began to take on water.

One of the sailors got hit pretty badly and I took some shrapnel in the leg but it was not too bad. And I did not have time to think about it because we were REALLY taking on water!

Two other sailors got out helmets and started bailing water like mad. We went to find a Red Cross ship but it was so dark we couldn't see our hands in front of our faces. With all the flares going up we finally found the hospital ship. In the mean time, we were trying to help the sailors bail. I wrapped my leg and did not think about it until we got to the hospital ship. One scene that is still in my mind was seeing a young Marine on a stretcher who had been hit. The bullet hit him in the head, passing behind his eyes from one side to the other. This man was simply talking to the corpsman as if it was just another day.

We thought that we might be able to stay on the hospital ship for a while but they hollered at us to move the Higgins boat since there were many more waiting to unload the wounded Marines and sailors from the landing. One of the corpsman asked me if I wanted my Purple Heart but before I could answer him an officer yelled at all of us to get the Higgins boat off the side of the ship or he was going to sink it and throw us over board. I was not hurt that badly but I now wish I had accepted the "heart" then. We did not go directly to our boat but went to find a mess hall since we had not eaten since day break. While we ate, someone was looking at our boat trying to get it repaired.

We had taken kapok life vests, tore them apart and stuffed the kapok inside the holes. By that time it was daylight so I told the boat driver to take me back to Iwo so I could find my tankers of the 5th Tank Bn. Well, I got ashore and before I knew it, some officer told me to go with him. I was only a PFC so I did what I was told. I was petrified since I had virtually no infantry training and here I was joining them in the field. They gave me an M-1 rifle and told me to shoot some Japs. On day two or three of the landing on Iwo Jima, I was with an officer. I was at his side when a Jap sniper began shooting at us. The officer aimed his rifle and shot the sniper out of the tree in one shot. The odd thing was that this Marine officer did not go over to see if the Jap was dead or not. He just kept moving. This was my first taste of combat. I was at #1 airstrip for two nights and I got to see them raise the second flag. What a great sight!

I did not like this infantry stuff at all. I was a tank repair man and there had to be some tanks in need of repair, so as soon as I could I went to find my tanks. To be honest, I don't remember what happened the next two days. I got some C rations from somewhere and had dug a hole in the sand (I called the sand on Iwo "coffee grounds" since that is what it looked like to me). For the next few days I found a lot of stuff to do. Help with a stretcher or help move ammo up the beach.

I recall taking chow up to the front lines. I know somehow I killed some Japs. Hell, you couldn't miss, they were everywhere. We always had two to a fox hole. We'd put the rifle between our legs so that if we were on watch and we fell asleep, our heads would hit our rifles and it would wake us up. They gave us small packages of cigarettes and maybe a small bottle of whiskey but since I neither smoked or drank, I traded mine for cokes. Finally I got to stay with my tanks.

My MOS was officially a tank mechanic but my training was with this secret gyro stabilizer., which was a contraption to help to keep the main gun of an M-4 Sherman tank steady on target, even if the tank is moving. Not all of the tanks had them but if one of the tanks that had one got knocked out, I had to get inside of the knocked out tank and pull this device out so that the Japs did not get ahold of it. My duty was to get the gyros from a stricken tank. How we would accomplish this recovery during the fighting was to pull a retriever or a working tank nose-to-nose with the knocked out tank. I would go out the escape hatch, crawl under the turret to grab the gyro stabilizer. Let me tell you, I had to crawl out of one tank's escape hatch, in through the knocked out tank's escape hatch and into the turret. When a Marine tank was knocked out, most, if not all, of the crew was usually killed or badly wounded. Some of the dead were left inside the tanks. It was dark inside the turret and usually covered with blood and gore. I would have to crawl around the floor of the tank and through all that mess. It was not a fun job. I wish I could have found the driver and the crew of that retriever and thank them for their bravery and for helping me do my job. Somewhere I cut my other leg again, but it wasn't that bad. I just wrapped it myself.

One day a Marine pickup truck came down the road with a bunch of Japs in the back, taking them to HQ for interrogation, when all of a sudden some Marines shot and killed all the prisoners right on the road. It was pretty awful. We also had one Marine who went a little batty. He would go around and knock the gold teeth out of the dead Japs mouths. Anytime he heard about some dead Jap somewhere he'd run over and find the gold teeth.

During the mopping up on the Northeast side of the island, I had heard that there was a Marine blade tank (with a full dozer blade on its front) that was firing at the Japs as they ran out of their caves. The Japs would jump into the water to try to swim away and the machine guns of the tanks would chatter. The tankers shot so fast and so furiously that the top part of the blade got shot up and it was all notched. I actually did not see the tank's blade but I was told about it many times.

I went to an Iwo Jima Marines' reunion in Biloxi, Mississippi a number of years ago. We were at a table talking about old times and absent buddies when a man pulled out a big roll of old Japanese money. He walked down the row of Iwo Jima Marines and handed each of us a bill from his wad. As he passed me, I told him that I knew where he got the loot. He looked at me funny and asked me to tell him. Well, a buddy of mine and I had gone looking around during the mopping up on the Northeast side of the island. We were forbidden to simply mosey around the island as we were doing but we went none-the-less. We came upon an old shot up building that had served the Japs as a store. So we went in and found a lot of Japanese money. All of a sudden something scared us so we threw down the money and bolted out of there. We did not want to get caught by a Marine officer. No sir. That could be worse than getting caught by a Jap!

I told him he had gotten the money from that store. He admitted it and here he was passing it around 65 years later. It's a small world.

M-4 Sherman used to haul wounded

Howard Dahms

Written by Clyde Hoch

On the island of Peleliu there is a bronze plaque near an old burned out Sherman tank, a remnant of WWII. It is so quiet and calm in contrast to the day it exploded and most of its crew killed. The Hell - confusion, death, maiming. One can only imagine what it must have been like for the men risking and giving their lives for their country so they could have a better way of life.

The tank was part of U S Army, 1st Platoon, "A" Company, 710 Tank Battalion. On October 18th 1944, the tank was ordered forward to support Marine units. Lieutenant Gilbert Lindloff was the commanding officer of the 710 Tank Battalion. He made the decision to keep the assistant driver in camp as he was one of the few that wasn't sick.

He felt he would need Charles Erazmos if another tank was called for up front.

The tank in which Howard Dahms was a gunner struck an Ariel bomb planted as a mine by the Japanese. The explosion was so powerful it punctured the hull of the tank. The tank commander John Prehm was blown off the tank. Assistant driver Private First Class George Lopes, driver T/4 Otto Hasselbarth and loader Corporal Michael Valentino were killed instantly, along with Marine Captain Howard W. Jones who was on the back of the tank directing the tank and his men. Capt. Jones was part of the 7th Marine Regiment of the 1st Marine Division.

Private First Class Howard Dahms was badly burned but still alive. He was taken to a field hospital where he later succumbed to his wounds on the 28th of October.

Howard was born and raised in Philadelphia, Pennsylvania. His father was a Fairmont Park Guard. Howard was fortunate to have grown up along the wooded Penny Pack Creek, where he had room to roam and play in the creek and woods.

Howard was one of eight children. He had two brothers, George and John, who also served in the military. George entered the Army at the end of WWII. John was an infant when his brother Howard was killed. John entered the Air Force where he served in Europe as a flight line driver. Howard had two nephews, Charles and George, who served in Vietnam, one as a sniper the other as infantry.

Howard Dahms along the Penny Pack Creek

Howard was one of the many who sacrificed so very much for the United States. It is my hope that his name will never be forgotten.

The battle for Peleiu lasted from September 15th until November 27th, 1944. Marine Major General William Rupertus was in charge of the American invasion forces of combined Marine and Army, with almost 17,500 Marines from the 1st Marine Division and almost 11,000 Army soldiers from the 81st Infantry Division.

The Japanese were led by Colonel Kunio Nakagawa of the 14th Infantry Division, with almost 11,000 troops. The Japanese had the advantage of 500 limestone caves. Some were old mining caves and some were natural caves which made ideal defensive positions. The Japanese had plenty of time to build up great defensive position. They were there to keep the island in Japanese hands.

The 1st Marine Division had 1,252 killed in action (KIA) and 5,274 wounded in action (WIA). The 81st Infantry Division had 542 KIA and 2,736 WIA. The Japanese had 10,695 KIA and 202 taken as prisoners.

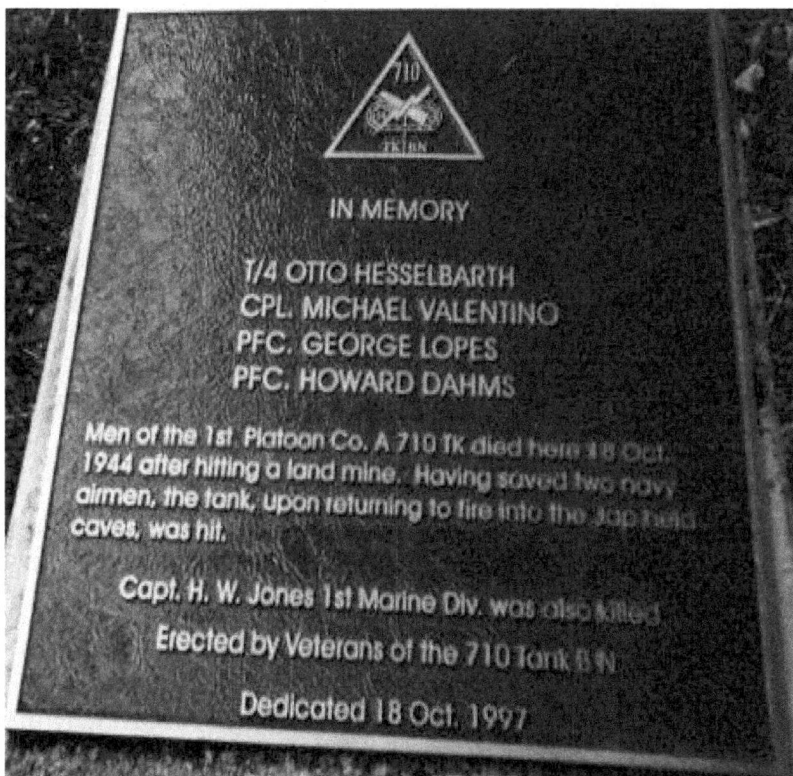

This plaque was dedicated on October 18, 1997
by the Veterans of the 710 Tank Battalion

Tanks of the Korean War
Written by Clyde Hoch

M-4's were used in Korea. Some were fitted with flame throwers. Some had 105 millimeter cannons. Some had a blade and were used as bulldozers.

The M-26 Pershing was used in Korea; it was a Medium tank, produced from 1945 into the 1950s. It weighed in at 46 tons. It was 20 feet 9 inches long and 11 feet 6 inches wide. It had a 90 millimeter main gun carrying 70 rounds. It had two 30.06 machine guns and a 50 caliber machine gun. It was powered by an eight cylinder gasoline engine that put out 450 to 500 horse power. Top speed was 25 miles per hour.

Later in the Korean War the M-46 Patton was introduced. It was a medium tank. It was produced from 1950 to the later part of the 1950s. It weighed in at 48.5 tons. It was 27.82 feet long. It was 11.52 feet wide. It had a crew of 5, Tank commander, gunner two drivers and a loader. It had a 90 millimeter main gun carrying 70 rounds. It had a 50 caliber machine gun and two 30 caliber machine guns. It was powered by a twin turbo gasoline engine, which produced 810 horse power. It had a range of 80 miles and top speed was 30 miles per hour.

Later still the M-47 Patton. It weighed in at 44 tons. It was 27 feet 11 inches long. It was 11 feet 6 inches. It carried a 5 man crew. Tank Commander, gunner two drivers and a loader. It had a 90 millimeter main gun and carried 70 rounds. It had two 30 caliber machine guns. It was powered by a Continental, air cooled, twin turbo gasoline engine. It had a range of 100 miles and a top speed of 37 miles per hour.

M-47 Patton

Major Roger Chaput
Written by Clyde Hoch
Photos supplied by Major Chaput

Pictured here as a Private First Class Chaput,
the Loader during the Korean War.

PFC Roger Chaput as a driver during the Korean War

Pfc. Hector H. BELTRAN   Pfc. Roger U. CHAPUT

Second Lieutenant Chaput June 3rd 1954.
The same day his second daughter was born.

Major Chaput was born in Nashua, New Hampshire in 1932. He graduated high school in 1949 and entered the USMC February 22, 1950. He served as an MP in 2nd H-3-1 before being sent to 3rd Platoon, Alpha Company, 1st Tank Battalion in Korea. He was wounded on June 3, 1951. He did not turn himself into the "A" Company Corpsman until June 4th. He became the first Corporal in modern Marine Corps history to become a recruiter. He was promoted to Staff Sergeant before attending Officers Candidate School.

Major Chaput served in many different rolls in the Marines including the 8th Communications Battalion, Motor Transport and maintenance battalions. Especially as loader on A-31 in Korea. He served on a ship in the Mediterranean and served at North Camp Fuji, Japan, the Philippine Islands to many duty stations in the United States. He served in Vietnam in 1966 and 1967 in motor transport. He held the rank of Warrant Officer 1 and W-2. Major Chaput retired from the Marine Corps in 1970 after a long and successful carrier.

His decorations include a Navy Commendation Medal with the combat "V", Purple Heart, Good Conduct with Star. Korean Service with 6 stars, and Vietnam Service with 3 stars.

Loaders Lament
Written By Major Roger Chaput

By 20 February 1951 winter had begun to ease up along the MLR 3rd Plt. "A" Co. led by 2nd Lt. G.G. Sweet, TSgt. Joe Sleger, Plt. Sgt. and Section Leaders SSgt Gerald Swinicke and designated point man SSgt Cecil Fullerton. At this point in the "Police Action" Russian built T-34 tanks were being knocked-out by Corsairs and never made it to the MLR. This allowed us to focus on tank infantry support for our Marines.

Now having turned 19 and being a seasoned loader on A-31, G.G. Sweet's tank, and being attached to the 1st Marines seemed like a great opportunity to bust a few more caps. G. G. Sweet and our gunner CPL. Capparazo made sure there were enough targets to go around.

All I remember is our platoon moved out and we buttoned-up. As the loader I couldn't see anything in front, behind or on our flanks. We moved through a narrow valley and suddenly I was too busy to see where we were going. I could tell we were beginning to climb as it took an extra hard shove to get the 90mm rounds chambered/loaded. We were going up some steep inclines.

I don't recall how many rounds of 90mm or boxes of 30 caliber machine gun ammo we expended that day, but I do recall that my eyebrows had been reduced to ashes and by the time we got unbuttoned, I was barely able to pull the bolt on the 30 caliber MG back with both hands. I strained with both arms, with all I could muster to charge the gun, but I did it! Lt. Sweet was calmly barking orders over the radio and Capparazo kept me jumping for the 90 HE, 30 caliber setting delays on the HE fuzes and more dammed 30 caliber ammo.

Thank God the 1st Marines reached their objective, when it was over I could barely stand up. I knew it was my job to replenish the ammunition in the turret. I was so gassed from gun powder smoke, I couldn't lift a box of 30 caliber ammo. Tank Commander/driver Joe Moreno and Asst. Driver/bow gunner Cpl. Dorman and Capparazo our gunner cleaned out the 90m and the 30 caliber shell casings still in the turret, replenished all of the ammo and fueled the tank with gasoline for the next day's operations.

Fifty years have gone by since then and I've had major surgery more than once, but I've never been so weak as I was at the end of that day. Thanks to the rest of the crew on A-31 when the sun rose the next day I was ready to bust a few more caps.

Working on the Railroad
Written by Major Chaput

Having enjoyed R&R in our pyramid tents and "B" rations in Masan, South Korea over the Christmas holidays, we mounted our tanks on railroad flat cars. The M-26 tank tracks hung over the edge of the narrow flat cars. As we climbed northward in a never ending tunnel, suddenly the train came to a halt halfway through the tunnel. Someone got off and walked toward the engine to recon the delay. As I recall the train consisted of 2 box cars carrying mucho doggies and 5 flat cars hauling 3rd Plt. "A" Co. tanks. The Platoon Leader was G. G. Sweet and the Platoon Sergeant was Joe Sleger. It seems the train's engineer and most of the doggies had passed out from the smoke and gas fumes from the coal-burning engine. By now steam is hanging from the tunnels overhead and as it cooled, it turned into a mild drip drip rain inside the tunnel. Outside the tunnel it was freezing, snow on the ground, inside it was a steam bath.

Someone figured out how to get the train into reverse and we backed down and out of the tunnel. The engineer was replaced, the doggies evacuated.

The humidity from the steam had taken its toll, the breach blocks on our 90mm guns turned from shining steel to rust. Cpl Capparazo, the gunner and myself on A-31 had the pleasant task of getting our weapons including the 50 caliber sky mount cleaned and oiled. Sgt Joe Moreno, driver/tank commander and Cpl Dorman, asst. driver cleaned up the bow gun and checked out the engine compartment. We made it to our destination somewhere near Wonju and ultimately back to the MLR. (Marine Logistic Regiment).

A-34 after hitting a mine in 1950

Y-51 mechanical problems in Inchon 1950

Tanks at Chinju 1950

Pusan perimeter

Chinese prisoners Chosin

3rd Platon, 1st Tank Battalion- North East Ma-Jong Dong, North Korea.
2nd row standing, far left Second Lieutenant G. G. Sweet.
Far right TSGt Sleger, kneeling far right PFC Chaput

Colonel Chesty Puller

Captain Granville G. Sweet
Written by Clyde Hoch

Captain Sweet was born in 1918, in Glenwood, Illinois. He served in World War II as a gun captain on the USS Nevada. During the Pearl Harbor invasion, on December 7th 1941, he was blown into the water and severely burned. Due to his injuries the Marine Corps decided to have him sell war bonds. He convinced his superiors that he was fit and able to continue to serve with the best of them.

He was transferred to 3rd Tank Battalion where he participated in the battle of Boganville and Peleliu. He was promoted to First Sergeant for the battle for Guam and Iwo Jima.

While en-route to Iwo Jima aboard the LST 477 (landing ship tank) the ship was struck on the starboard side, by a Japanese Kamikaze pilot. It was here the Captain (then Sergeant) earned his second Navy/Marine Commendation

Medal. At the end of WWII he was commissioned a Second Lieutenant.

During the Korean War he was a platoon commander of 3rd Platoon, "A" company 1st Tank Battalion, 1st Marine Division. In 1951, he earned a Silver Star for outstanding leadership. The platoon he was commanding stopped the advance of the North Korean People's Army's 7th Tank Regiment at the first battle of the Naktong River, west of Pusan, South Korea.

Captain Sweet spent most of his Marine Corps career in tanks. After he retired from the Marine Corps he started a memorial park and museum in Pahrump, Nevada. He wanted to honor the people who served him so well. Names of people who touched his heart are carved in stone.

Included in the names is Ray Kroc. Ray was an Army Veteran who convinced the Captain to purchase one of the earliest of McDonald's franchises. The Captain went on to purchase a chain of McDonald's restaurants.

Major Chaput presided over Captain Sweet's funeral. They were good friends and the Major was with Captain Sweet to the end on July 5th 2010.

Lt. Sweet gets the Silver Star in Quantico, Va.

THE LIEUTENANT
1918-2010

DLA-rdw
043134-2

Capt. G.G. Sweet USMC
1918 - 2010

Spot
1stMarDiv
Ser 35418

The President of the United States takes pleasure in presenting
the SILVER STAR MEDAL to

SECOND LIEUTENANT GRANVILLE G. SWEET
UNITED STATES MARINE CORPS

for service as set forth in the following

CITATION:

"For conspicuous gallantry and intrepidity as a Platoon Commander in Company A, First Tank Battalion, First Marine Division (Reinforced) in action against enemy aggressor forces in Korea on 23 February 1951. Assigned the mission of supporting an infantry battalion in the attack on a strongly defended enemy hill position, Second Lieutenant Sweet fearlessly and with complete disregard for his own safety exposed himself to devastating enemy fire, skillfully leading his platoon forward over the tortuous terrain seeking routes of advance which would enable the platoon to move in close support of the infantry. Repeatedly moving on foot to better direct the fire and movement of his tanks, he was able to keep his platoon with the assault elements, rendering effective close-in machine-gun fire, and destroying bunkers at point blank range. His aggressive leadership and outstanding devotion to duty were an inspiration to all who observed him, and aided immeasurably in the successful seizure of the objective. Second Lieutenant Sweet's heroic actions were in keeping with the highest traditions of the United States Naval Service."

For the President,

Signed
6-4-63

FRED KORTH

Secretary of the Navy

R. Glenwood, Illinois
B. Glenwood, Illinois
16 Jul 1918

SecNav This Copy

Sweet's certificate for a Silver Star, which he hid from his men because he felt he didn't deserve it.

Monument at Sweet's Park and Museum

Spring offensive 1951. Photo supplied by Major Chaput

Spring offensive 1951

Colonel Joseph Sleger Jr
Written by Clyde Hoch

No book about tanks involving Korea or Vietnam would be complete without a short dedication to Colonel Sleger. The Colonel was born in 1927 in Green Bay Wisconsin. He entered the Marine Corps in 1945 and went to boot camp in Parris Island, South Carolina. Private Sleger was sent to a weapons company in Tsingtao, China and here later served with 3rd Tank Battalion 4th Marines. Now Sergeant Sleger was transferred to "A" company 1st Tank Battalion Camp Del Mar, California. In 1950 Sergeant Sleger was transferred to 1st Marine Brigade in Korea. Here he took part in countless operations against the Chinese and North Koreans, where he was highly decorated.

He was transferred to instructor at the tank schools at Camp Del Mar and was promoted to Master Sergeant. In 1952 he was commissioned to Second Lieutenant and served as a company commander in "A" company 3rd Tank Battalion. He also served as Combat Cargo Officer aboard the Attack Transport Ship the USS Magoffin to assist in removing refuges and French Forces from Haiphong, North Vietnam during the French Indo-China War.

He became a Company Commander for the Infantry Training Regiment in Camp Pendleton, California. Later a Guard Officer and Executive Officer at Marine Barracks' at the Naval Torpedo station at Keyport, Washington.

Now Captain Sleger became a Company Commander for the 2nd Tank Battalion at Camp Lejeune, North Carolina. Next the Associate Armor Officer course at Fort Knox, Kentucky.

He became a company commander in 3rd Tank Battalion in Okinawa, which covered combat operations in South Vietnam including the Mekong Delta.

Major Sleger became the Executive Officer of the Recruit Training Regiment in California. He attended the Marine Corps Command and Staff College in Quantico, Virginia.

Coloniel Sleger became the G-4 Operations Officer for the 3rd Marine Division in South Vietnam and later the Commanding Officer of 3rd Tank Battalion in Vietnam.

He retired from the Marine Corps after a long and dedicated career on July 1st, 1978 and passed away on July 8th of 2012.

Although I never met the man, I know he had a great influence on my tank years and that of tankers in today's

49

military. His memory and guidance still exist today and will continue for many years to come.

Look at this platoon. Made up of LT. GG Sweet, Sgt. Joe Sleger and PFC Roger Chaput. Imagine all the talent in this platoon. Lt Sweet was an enlisted man who went on to become a Captain. Joe Sleger was an enlisted man who went on to become a Colonel and Roger Chaput was an enlisted man who went on to become a Major. These men were respectfully called Mustangs. A Mustang is an enlisted man who becomes an officer. They had more respect than anyone in the Marines, because they worked their way up from the bottom.

If you served in tanks yesterday, today or tomorrow, Army or Marines you must certainly have been influenced by these great tank commanders. It is my honor to serve them by including their stories in my book.

TSGT Sleger (Center) January 1951

Tanks of Vietnam
Written by Clyde Hoch

The most widely used tank of Vietnam was the M-48A3 or Patton. The name Patton was given too many tank designs, with a large amount of variations, starting with the M-47. Including the M-67 flame tank or Zippo as it was sometimes called.

The M-48A3 was a medium tank weighing in at about 50 tons, 52 tons combat loaded. It was about 31 feet long, 12 feet wide and carried a crew of four.

It was armed with a 90 millimeter main gun and carried 60 rounds. It had a 50 caliber machine gun. Some of the 50's were mounted inside the cupola and some on top of the turret. The inside of the cupola way of mounting had a limited amount of movement, caused jam ups and limited visibility. Some were mounted on top of the cupola and called a sky mount. The disadvantage of the sky mount was when it was fired the person firing it was exposed. It carried a 30 caliber machine gun that was mounted along side of the main gun. The gunner had the choice of the main gun or the 30 caliber machine gun.

The M-48A5's were armed with a 105 millimeter main gun, this variation came out in the mid 1970's.

The M-48's were originally powered by a Continental Air cooled twin turbo V-12 gasoline engine that put out about 810 horse power.

The M-48A2C was powered by a Continental air cooled Twin Turbo V-12 diesel that put out about 750 horse power. They had a range of about 300 miles and a top speed of 30 miles per hour.

The M-67 flame tank was used in Vietnam. It was very effective for heavy bunkers and fortified positions. Built on an M-48 Chassis. The M-67 was almost always accompanied by gun tanks.

M-48A3 or Patton tank the main tank of Vietnam

Amphibious Tractor LVTP-5

Written by Clyde Hoch

In the following stories we talk about AmTracs (amphibious tractors). AmTracs were a large amphibious personnel carrier. They could carry from 34 to possibly 40 troops. I was told the engineers said this monster would not float, but it did. Very little of it was exposed while it was in water.

We escorted AmTracs on many occasions. They had thin armor and a gas engine, therefore they would burst into flames if hit at the right spot by small arms fire or if they hit a mine. They were powered by a 12 cylinder gas engine.

AmTracs had a crew of three - a commander, driver and machine gunner. They weighed in at 39.9 tons, 29 feet 8 inches long, 11 feet 8 inches wide and ten feet high with a top speed of 30 miles per hour on land 6.8 miles per hour on water. They had a range of 190 miles on land and 57 miles on water.

AmTrac on Land

AmTrac in Water

M-50 Ontos

All of the Ontos photos were supplied by Corporal Pat Tierney, who was used as an artist when needed. He served in the 1st Anti-Tank Battalion. Story by Clyde Hoch

Ontos were called "Pigs" by most Marines. They were originally used as anti-tank weapons but quickly became useful as direct fire support for grunts. It weighed in at about six tons, carried a crew of three. It was 12 feet long and eight and a half feet wide.

Ontos firing its six 106 milimeter guns

The Ontos was a small, light and could sometimes travel rice paddies where heavier weight tracked vehicles couldn't go. The Ontos had six recoilless rifles, with four 50 caliber rifles for spotters. It also carried a 30 caliber machine gun. Draw backs to the Ontos were the crew had to get out of the vehicle to reload the 106 rifles. They could only carry 18 rounds of ammunition for the main gun. Ontos were powered by a six-cylinder engine.

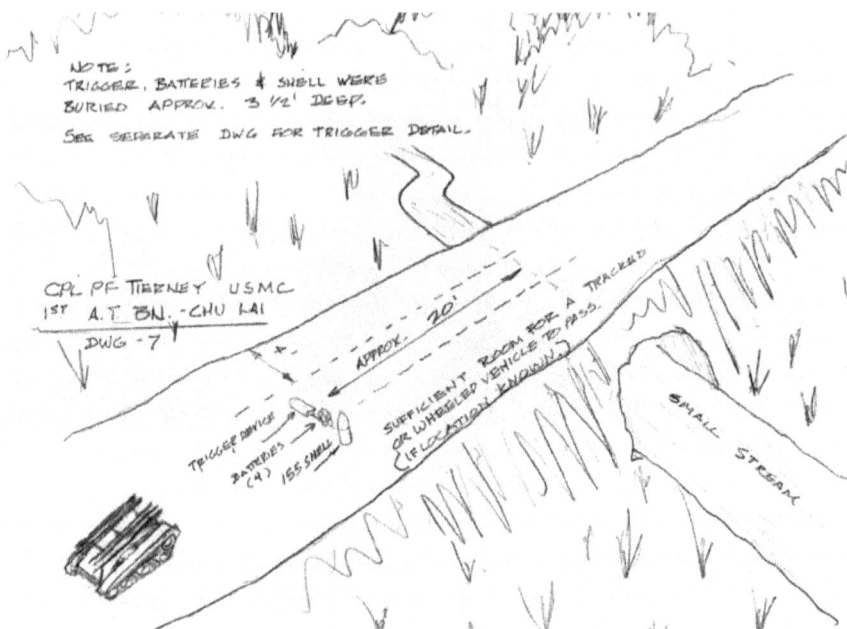

Drawing by Pat Tierney

The drawing is a description of an Ontos hitting a mine. The mine was set at a depth where bicycles and motorcycles could pass over it but the weight of an Ontos or heavier vehicle would detonate the mine. The driver of the Ontos was killed. He had ten days remaining before his release from active duty. The rest of the crew were wounded but survived.

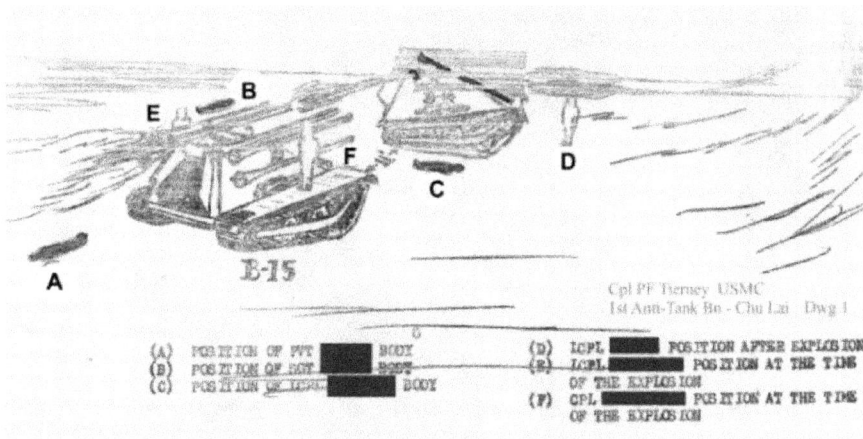

Cpl PF Tierney USMC
1st Anti-Tank Bn - Chu Lai   Dwg 1

(A) POSITION OF PVT ███████ BODY
(B) POSITION OF SGT ███████ BODY
(C) POSITION OF LCPL ███████ BODY

(D) LCPL ███████ POSITION AFTER EXPLOSION
(E) LCPL ███████ POSITION AT THE TIME OF THE EXPLOSION
(F) CPL ███████ POSITION AT THE TIME OF THE EXPLOSION

An official document of an accidental explosion. It is felt the loader thought he unloaded all the 106 millimeter guns. He missed one and slammed the breech shut causing it to fire. The back blast killed him. The round hit the guns on the next Ontos and two of them fired, just missing the ammo Dump.

This photo was taken in 1969 south of Danang. Standing left to right, Clyde Hoch, Richard (Ski) Gerszewski . Kneeling left to right Jerry Holly and Todd Phillips. The photo was taken shortly before I rotated back to the States.

These were great guys. I was proud to have served with them. Todd and I are the only two surviving in this photo. Todd and I keep in touch often. I still call Todd a very good friend and brother after all those years.

Todd Phillips

Written by Todd Phillips

This is going to be a short story because there are a lot of things that I don't remember about my childhood and Vietnam. Some psychiatrist might be able to figure it out, but I think it might be best to leave it alone. I can live without the memories, so we'll just leave that box closed. I have a couple of friends that can remember everything that they ever did, along with names and dates. Not me. I'm really bad with that.

John Wayne had a code. He wouldn't be wronged, he wouldn't be lied to and he stood up for the things that he believed in. I still think he was right. Being a man has its responsibility. I think that defending this country is at the top of the list. I get so sick of the cry babies carrying signs and complaining about our country. No it's not perfect, but if we don't stand up and fight for our rights, they wouldn't be able to carry their signs and they would lose their right to free speech. Then what are they going to do? I think we owe a debt to our fathers and fore fathers for what they have done for us. They gave up a lot. Some gave everything so that we could have the rights that we have today.

Childhood
Written by Todd Phillips

They tell me my mother was a wonderful woman. Maybe so. I couldn't say. I can't remember back that far. I've only known the woman for a little over 60 years. Like Forrest Gump "that's all I'm going to say about that."

As far as my father goes, I wish that I had been a better son, but being a stupid kid, I thought he was a weak man and I didn't have very much respect for him. What is it they say, "My father didn't know a thing when I went into the service but it's amazing what he learned in a few short years." The truth be known, my father was the kindest, smartest and the most gentle man I have ever known! I took this for weakness, but oh how wrong I was. To know him, you would never know that he was a decorated "SILVER STAR" soldier, serving in the Philippines in WWII, all I know is that I love, miss and respect him very much!

Todd's parents Freda and Garland Phillips

High School
Written by Todd Phillips

High school was a bit of a blur to me. I never cared about school work. I don't think that I took a book home to study in all four years of high school. I always knew I was going to make a living with my back and not my brain. I was always good with my hands and practical math. They served me well in the construction industry. Spelling and English not so much.

I came from a small town in East Central Ohio, on the river. It was a great place to grow up. Everyone knew everybody. Sometimes, though, that's not a good thing. You old neighbors get to tell your kids everything they did when they were your age. The town had about 15,000 people at its peak. Today it's less than 5,000. It is no longer a city, it's a village. The steel mills, the coal mines and the glass plants are a thing of the past. When I was a kid, people had pride in their homes. Even though it was an old town, people took care of their property. Now, in this economy, the town is falling down. It is a shame.

Even growing up there in the 50's and 60's I can honestly say I don't remember there ever being a race problem. Maybe it was due to my ignorance, but I don't remember any. Martin Luther King was murdered on April 4th 1968 and not long after I was gone.

USMC
Written by Todd Phillips

I turned 18 on 6/2/68, graduated on 6/6/68, enlisted in the Marine Corps on 6/10/68 and was in Parris Island on 6/19/68. I made it to Nam right after New Years. Just think, in six months I went from being a high school kid to a lean mean fighting machine, or so they told me. People asked me where I went to college. I tell them I went to the University of I Corps. What is it they say? "The only Woodstock that I remember in 1969 was on a M-14".

When I think about it, there were a lot of things we missed in 1969. We missed the aftermath of the killing of MLK (4/4/68) and RFK (6/6/68). The moon landing (7/20/69), Woodstock (8/15/69), and the Charles Manson murders, just to name a few. Even Kent State. I'd been home less than 4 months from the Nam when that took place so I hardly remember it at all. That time was just a blur in my mind.

I have a good friend who told me once that our whole world had changed in one short year. When I think about it he was right. Not so much from the effect that the war had on us, though it did have an effect and still does. Just think about it when we left home, America was still kind of normal, but oh my God, what the hell happened while we were gone? The drugs, the music, the Hippies and all the protesters. They were protesting US. US as in me and US as in the United States of America. I couldn't believe it. Here we were doing what we thought was right and what was expected of us by our families, our country and our community. Hell, none of us wanted to go, but it was the right thing to do. I've had family that has fought in damned near every war that this country has had from the Revolutionary War, Civil War, WWI, WWII, Vietnam (me). I didn't forget Korea; it's just that I don't think anyone was at the right age. They were either too young or had just fought in WWII. Come to think of it, I can't remember any male, except for my uncle in my family that did not serve his country. We had Army, Navy, Air Force and one MARINE!

My grandmother said that she felt like she didn't do her part. It was on her side of the family that fought in the Revolutionary and Civil Wars. Her husband, My Grandfather, was a veteran of WWI. She had two brothers in WWII, a son, a daughter and son-in-law that were in WWII. Her one brother, my Uncle Clyde and a son, my Uncle Forrest were in the battle of the Coral Sea (I think) her son-in-law, my father, was in the Aleutians and the Philippines. Later he entered Japan with the 25th Inf. Div. She even had a grandson in Vietnam, me! I think that she did her part. She was a great Grandmother and wonderful person.

Boot camp was one hell of an experience. Did you ever wonder what a fish feels like when someone reaches down and pulls him out of the water? Out of the only world it knows, into the unknown? Well that's kind of what I felt like when I got to boot camp. My whole world had changed. Like most guys I spent the first few days hoping to wake up and be back home in my nice, safe warm bed.

When we first arrived at MCRDPI (Marine Corps Recruit Depot Parris Island), this drill instructor came on the bus and started hollering and screaming at us. He told us to get on the yellow foot prints. Anyone who was there knows what I'm talking about. Well I was the first one off the bus and I went over and stood on the yellow foot prints. Some guy came out of the barracks and told me to get in there. I tried to explain to him that the man told me to stand here. Well then he started hollering and screaming. Since I did not want to cause any trouble, especially my first day, I started to walk up to where he was at. That's when he grabbed me by the back of my neck and started kneeing me in the ass saying "you don't walk around here, you run!" Well not being used to being treated like that, I turned around and got ahold of him. About that time he had three or four friends that decided to help him. I'm not the sharpest tool in the shed, but I learned real quickly not to be the first at anything. You just can't do anything right in boot camp. Boot camp is what it is. You either like it or you don't, but I will say that they did more to turn the boy into a man in twelve weeks, quicker than anything else had in the prior 18 years.

I would like to make one suggestion, if you're going to join the USMC and go to Parris Island, do it in the fall, winter or spring, because it's hot in June, July and August! When graduating from boot camp, I was told my MOS was 1800. Nobody had a clue what an 1800 was. After some checking, we found out an 1800 was tracked vehicles. In the USMC everyone from cooks to pogs (clerks) to the Commandant has an MOS of 0311 (basic infantry). Then you get a secondary MOS. Mine ended up being 1811 which was tanks. I didn't think much about it then, but looking back, I couldn't believe how lucky I was to get into tanks. The big pig was a big target, you couldn't hide it in a fire fight and you were your own 52 ton mine sweep, but it beat the hell out of a flak jacket. I really don't think that I'd have made it back from Nam if I had to hump in the boonies!

I remember when I enlisted. On the paper work they asked you to write down the three jobs that you would like to do. It's not because you would get one of them. Thank God. I wrote down, machine gunner, gunner on a chopper or infantry. I bet when they read that they said, "This dumb ass hasn't got a clue," and they were right.

Going to Nam
Written by Todd Phillips

When I was in boot camp and ITR (infantry training regiment), I knew everyone that I was with. Then in schools and staging, I knew fewer guys. Well, when I got to Okinawa, I didn't know anyone. On the night before going to Nam we got a little drunk, and it was raining like hell. The Sergeant said that there was no reason for us to get all wet, so we were told to get on the bus and catch some Z's until it was time to go to the airstrip. When we got to the airstrip, we started loading by name and number. When they were finished, I was still standing there. The troop handler asked who I was. I told him, and he informed me that I had gotten with the wrong group. My group had already left and I was AWOL in a war zone. Not a good thing! I was told to meet with the base commander at daybreak. I was not looking forward to this at all. Well, I got lucky. I got a flight out with officers and corpsman. Now, here I was a slick sleeve private on a plane with all this big brass. When the sun came up, I was almost blinded by the reflection coming off of all the gold and silver.

Like most guys the first thing that hit me getting off the plane was the heat and smell. I guess I was so scared on the flight over that I didn't think what might lie ahead in a war zone.

I was just standing around looking like a new turd in a big shit hole, when a jeep pulled up and asked if there was anyone for tanks. That's when my luck turned for the better. Little did I know that I was about to form a friendship that would last for over 40 years. One of the guys in the jeep was a tank commander (TC). His name was SGT. Clyde Dennis Hoch, but nobody and I mean nobody called him Clyde. Denny was OK, but most of us called him Sarge. I still do to this day. Like I said my luck had changed. Sgt. Hoch put me on his tank C-35. I'm glad that he did. Not only was he a good TC, but he was very knowledgeable about the tank and tactics. He knew what to do in a fire fight. I think that the best thing about him was he knew how to handle and treat the men around him. We did what he asked of us, out of respect, because he earned it. He was my tank commander 42 years ago and I still respect the Sgt. and the man. Most of all, I'm glad to still have him as a friend and brother.

When it comes to Nam, I'm not going into a lot of detail, but I will tell you about a few things that still seem to creep into my dreams 42 years later.

One was a pretty little girl. She was wearing her little white silk outfit riding a bicycle, probably on her way to school. I say little, but she was probably 15 or 16 years old. She was riding in the opposite direction. We were running somewhere around 25 MPH. For some reason, probably because I was only 18 years old, I turned around to look at her. We had picked up some razor wire in our rear sprocket and we caught the girl in it and was dragging her behind us.

We probably dragged her for a hundred yards before we could stop. The poor little girl was hollering and crying. I felt so sorry for her. Her little white dress wasn't so white anymore and I don't think that there was a square inch of her that wasn't cut and bleeding.

There are incidents that I wish I could go back and change, but since that will never happen; I have to live with it.

We were out on some no name operation or sweep. We were out all of the time. I don't think I spent more than a week back in the company area during my whole tour. We were running low on fuel, so they brought fuel out to us by chopper in 55 gal drums. We had a hand pump on the tank to transfer fuel. I think they called it a centrifuge pump, but ours didn't work. We had to pick up the drums and put them on the back of the armor plate (engine cover) so we could pour the diesel in to the fuel tank.

We had a drum sitting on the back of the tank when the grunts hollered "Fire in the hole". They had found a 500 lb bomb that had not exploded and were going to blow it in place. They did this to keep the Gooks from getting ahold of it and getting the explosives out of it or making a big tank killing mine. From the time that they hollered "fire in the hole" until we heard the explosion, there was only a split second. They didn't give us time to take cover. Ronny Lyons, the TC was down inside the tank so he was safe. The driver dove behind the drum and I dove off the side of the tank. After the dust cleared I climbed back up on the tank. The driver was holding on to his leg whining and rolling back and forth. Ronny stuck his head up out of the TC hatch and asked what was wrong with him. I told him the dumb ass had hurt himself when he dove behind the

68

barrel. You can't fall on a tank without hurting something. So I started kicking him, telling him to get up and calling him a pussy. That's when I noticed his boot was full of blood. There was a piece of shrapnel about an 1 ½ inches wide and maybe 4 inches long that had hit him in the shin bone and shoved it out the back of his calf muscle. And here I was kicking him and calling him a pussy.

Now, he was one of those guys that you meet and just don't like. We had got into a couple of fist fights prior to this. Now I know that I'm an asshole, I've been told this by a number of people, but I wouldn't treat a dog like that. I don't remember his name. I think that he was from the Midwest. I never saw him again after he was MedEvaced. I don't think he knew that I was kicking him and calling him names. He had to be in shock.

There was another time, another no name operation; the grunts tripped a booby trap. It killed their radio man and either their Sergeant or their Lieutenant. I was the Tank Commander on C-21 at the time. As C-21 goes, it was the first tank in the second platoon so when a LT wanted to go out, he always took my tank. When he took over my tank I went to the gunner position. After the grunts were killed, someone got on the radio and asked our LT. if we could hit the village on the other side of the river. He was excited and told them that we could. He told me to fire when ready. When you're out driving the tank around in rough conditions the constant bouncing would cause the gun to be knocked off of sight a little bit so the first round I fired which was an HE (high explosive) hit around 150 yards short, so I put the cross hairs on the point of impact and fired again. This round (another HE), hit about 30 yards short. So again I adjusted the sights. Then it got serious: 18 shots, 18 hooches. I never missed. When I stuck my head up to get some fresh air the grunts were cheering. It sounded like you were in Roman Colosseum.

It didn't bother me then, but looking back on it years later, I think it was senseless. I don't think they had anything to do with what happened, but you never know. Farmers by day, VC by night. But I often wonder how many kids, women and old people never even had a chance to get off their little bamboo mats and run. Since I can't go back in time and change things, I just have to live with the memories. Some nights I wish my brain had a switch on it so I could turn it off.

Ready to roll. Charlie Company headquarters,
South of Marble Mountain

The Rat Patrol
Written by Todd Phillips

It's funny how people remember the same incident and we all remember it a little differently. Over the years I've talked to some of the guys that were there that night and, like I said, we all remember it differently.

So this is how I remember it. We received orders to take our tanks out and run a rat patrol from the company area to a little village called Nui Kim Son. They ran rat patrols in order to keep the roads safe and supposedly free of mines. This is normally done with a jeep that has an M-60 machine gun mounted on the back. This works well because the jeep is fast, small and fairly quiet. Someone with all that shiny stuff on their shoulders had decided it would be safer to do it in a tank. Well you don't have to be a brain surgeon to know that you can hear a tank start up and run from miles away, especially at night. Come to think of it, trying to be stealthy in a tank would be like trying to sneak up on a deer with a Harley Davidson!

We made our first run without incident, but on the second we got our asses tore up. I was gunning and the first RPG (rocket propelled grenade) hit in front of me. The shrapnel hit the loader, Jim Littman, in the eye and hit the TC, Ronny Lyons, in the wrist. It hit the driver, Jerry Holly, in the head and neck causing him to drive off the road down into a rice paddy. He was knocked unconscious for a split second and he was blinded by the flash, he had blood in his eyes and even though he was screaming that he'd been hit, he was able to pull us back up on the road into a firing position. At that point, Ronny was returning fire with the 50 cal machine gun. At the time Jerry had keyed his mike open, so you could imagine what it sounded like. I can't put it into words. Ronny started kicking me in the back of the head telling me to fire, "Fire you're freezing! Fire".

Now this was my first major fire fight, proper procedure is for the TC to tell the gunner come left, come right, up, down. Now I know that when the shit gets that heavy the book gets thrown out. What Ronny didn't know, was all I could see in the scopes were muzzle flashes, and I didn't know if it was our guys on the other tank or our grunt security, and I sure didn't want to kill our own men. We were hit with a second RPG and a ton of small arms fire, so I started firing back. I think we fired about 28 rounds of 90 MM and 3 or 4 thousand rounds of 30 Cal. This all took place within a few yards of the tank. You could have reached out and shook their hand.

Now other tankers will know what I'm talking about, when you fire the 90 MM, when the breach opens it sucks the gasses back inside the tank. Back to the book again, when you fire a round, you throw out the spent casing and load another round, well that's bullshit because when things get that hot and heavy you don't have time to throw out the spent casings. When you get 20 or 30 of them on the floor they start binding up and moving against your legs as you traverse the main gun. I forgot to mention them SOB's are hot too. So now you have to picture this, with the gasses your eyes are burning and watering and you've got snot running out of your nose, your throat is on fire like you're back in the gas chamber and your legs are getting burnt from all the 90 MM brass, and all the while that this is happening, there is a lot of blood, people screaming, someone kicking you in the back of your head and oh yea, a bunch of little dinks trying to kill you and your friends.

There's more to tell about that night, but that's all I can stand right now. I will tell you that when it was over, I stayed on the guns to make sure we were safe while calling in the MedEvacs.

When I finally got a chance to come up for air, damn near everyone was gone. One minute they were there, then the next they were gone. At that point, I was told I'd have to drive the tank back in. Now I had not driven a tank since tank school and being a gunner you looked at Nam through a scope, so I wasn't sure where I was. Hell I didn't even know where the company area was for sure. After all it was very dark and you sure as hell didn't want to turn on the lights. On the way back I noticed I was wet and sticky. The driver's compartment is very tight and small. When we got back in, everyone wanted to know what happened. I guess when it happened, everyone in the company manned the berm to watch.

The captain said, "I told you guys not to take any shit, but I didn't tell you to start another war."

I kind of wished I could have seen what happened. I might have been a lot closer to the action, but all I saw was flashed through the scope. Guys started asking me where I was hit. I said that I wasn't hit. They said, "Yes you are." Hell, I was still numb. I could have been hit and not realized it. They had me see our Corpsman, Doc Forsyth. Then I saw why they thought that I was hit. The reason I was wet and sticky was that I had Jerry's blood all over me from being in the driver's seat. There are four men on a tank. That night on

C-35 they awarded three Purple Hearts and two Bronze Stars.

On a lighter note, I remember going to battalion for a Q (quarterly tank) service. Our BTN (battalion) sat on top of a mountain overlooking Da Nang air base. If you went to Dog Patch and Four corners and looked straight up you'd see us. When the gooks hit the airstrip with their 120 MM rockets, it was like watching a movie. Looking down on the airstrip you could see the rockets hit and watch the fire trucks go out. At BTN they had these hooches for us to stay in while they worked on our tanks. I think they called them South Seas Island hooches, but there was never any place to sleep. You took your poncho liner and grabbed a hunk of floor. BTN is where all the brain power is located. Come to think of it that's where SI (Supply One) was located. Someone got the bright idea to build us bunk beds out of 2x4's and plywood. They did a nice job, even nailed them to the floor - a row down both sides with a little walk way down the center. So we were all sleeping one night in our new bunk beds when Victor Charles decided to fire a few rockets at the airstrip. He had a couple of short rounds that hit the BTN area. Now I don't know if you've ever seen 10 Marines trying to get into a fox hole through a small walk way at the same time, we were using each other's ass holes for stepping stones. When it was all over, there was nothing left but a big pile of kindling!

Pipestone Canyon
Written by Todd Phillips

One of the biggest operations I was on was called Pipestone Canyon. The op began in mid March and ran through November. The op took place in Dodge City and Go Noi Island. The place had long been a strong hiding place for the NVA (North Vietnamese Army) and local VC. We not only cleared out the enemy, but it was also a land clearing operation to deny the enemy a hiding place. Dozers were brought in and cleared over 8,000 acres of land.

On the op, statistics gathered from "Semper FI Vietnam" said we killed 852 enemy, 58 captured, while Marines had 71 KIA and over 600 wounded. The place was a nightmare for booby traps.

Korean Marine Corps
Written by Todd Phillips

When I first got to Nam, I was on C-35, Sgt Hoch's tank. Our platoon was stationed at the South China Sea in support of the KMC (Korean Marine Corps). I must say that if not for a war going on, I could have built a Tiki Hut on the beach and lived there for the rest of my life. It was absolutely beautiful, palm trees, beach, and the sea. Wow. If I remember right, we had two tanks, maybe three at the platoon area and two tanks with the Koreans. We all laughed when we first saw the Korean compound. You never saw so many sand bags in your life, we called it the great wall of Korea.

Somehow, I don't know why, in late March or early April they split up our platoons. Ronny Lyons and I went from third platoon to second platoon. I never served in the same platoon with Jerry Holly again. Ronny and I were put on C-21. The name on the tank, which was on the gun tube was "Momma Cass" - all 52 tons of it. We were then rotated back to the Korean compound. After Ronny rotated back to the world, I was made TC on C-21.

Everyone had their own opinion about working with the Koreans. I'm not going to tell you how I felt about it, they've got enough trouble over there and I don't want to start another war.

A tank aint worth a damn without a support unit or a mine sweep. You've got a lot of fire power, but you're just one big target and you can't hide a tank. You are your own 52 ton mine sweep and some little gook will stick an RPG up your ass. Well, we were out in the middle of nowhere and the sun started to go down. All of a sudden we noticed no mine sweep in front of us and no support on our flank. They were all gathered behind the tank. When we tried to get them back out, all we got in response was "boo-koo boo-koo VC, no sweep, too dangerous". So what are you going to do when your ass is already out there? When it got too dark to see, I had to get off and ground guide my tank. We were going over this little hill and from our left side, an Amtrak (personnel carrier) was crossing close and in front of us. He must have been 30 feet away when he hit a mine. I had stopped my tank and was standing in front of it. I was between the tank and the Amtrak. The concussion from the explosion knocked me down. I was dazed and couldn't hear anything for a short while. My loader (can't remember his name) was hit in the eye with shrapnel. What happened to my driver still amazes me. He had stood up in the driver's compartment and the concussion from the explosion hit him in front of his thigh. It ripped the skin away and I guess the pressure from the blast pushed all the blood back, because looking at his leg was like looking at a diagram of the human body in the encyclopedia. You could see muscles working and the veins, but no blood, at least at first. I don't know how many men were killed in and on the Amtrak; it was full inside and on top. Being powered by gasoline, that metal coffin was all aflame. I don't know if having a mine sweep out in front would have kept this from happening, but we'll never know because they were all hiding behind the tank. At least when you were out with other United States Marines, everyone did their job and every one watched each other's back.

Going Home
Written by Todd Phillips

It was January 2, 1970. I had just returned to LZ Baldy after running a convoy up to LZ Ross, which I was not too happy to go out on. I was short. I mean I was so short that I had to look up to see a piss ant's belly. After getting back in, I went to see the head "shitter" which was located back by the comm. center (communications). As I sat there I heard that radio say tell LCPL (lance corporal) Kilo and CPL (corporal) Papa to get their sierra together, their flight back to the world was leaving in a few hours. By the time they came to tell me, I was already packed and ready. It was late and we were a long way from battalion. By the time we got there SI was closed. We threw our 782 gear and our MDL 1911 A1 45's on the floor, grabbed our orders and hauled ass for Danang airstrip. We barely made it. I sure didn't want to miss the great big beautiful freedom bird. We flew to Okinawa for a couple of days than to Camp Pendleton for discharge on January 10, 1970.

It's hard to explain this to someone who didn't go through it. On January 2nd I was in a war and eight days later I was walking down the main street in my home town a civilian. Man, talk about not being ready. In some ways it was as scary as going to Nam. Jody had got my girl and she was gone, no job, most of the people that I had graduated from high school with were either in college or in the service. What had happened to the world I'd left a short eighteen months and twenty-one days ago? Long hair, drugs, the music, the way people looked at you. They either wanted to talk about shit that you didn't, or they just kept their distance.

Todd and Brother Keith

The first job I got was working in the coal mines, the old way, setting post and shoveling. It was a hard and dangerous job, but it paid good. I was married, had a son and divorced in less than one year. I was having a lot of trouble fitting in. The mines went on strike, so being out of work again, I decided to go to Wisconsin and see my buddy Jerry.

Todd and Jerry in Vietnam

His dad made a couple of phone calls and I got a job at Snap On Tools. It was in the foundry working on a trimming machine. Man, I just couldn't see myself standing there in front of that machine doing the same thing as fast as you can day in and day out for the rest of my life. Jerry was the only person that I knew, so I felt alone and out of place once again. It got to the point that I wished that I had never left the Marine Corps. I went so far as to see a recruiter and take a physical. All I had to do was raise my right hand again, and I would have been back in. The only reason I didn't go is that they would have dropped me a rank, and they couldn't guarantee me that I wouldn't go to Nam again. Once was enough.

In Wisconsin, winter comes early. It started snowing in the middle of October. That's when I figured I had enough. I told Snap On that they could keep my check or mail it to me in Florida.

Before I was 21 years old, I had left home, gone into the Marine Corps, gone through a war, married, had a son, divorced and I was on my own again. Seems like that's the way my life has gone for the past 42 years. I just couldn't find a place where I really fit in.

I spent most of my life installing vinyl siding and roofing, I liked it. You could work on someone's house for a week or two, finish it and go somewhere else. I worked my own hours, didn't have to punch a clock and was my own boss. Talk about not staying in one place too long. I worked in

Ohio, West Virginia, North and South Carolina, Georgia, Florida, Indiana, Wisconsin, California and Vermont.

It's hard for me to put feelings into words: loneliness, emptiness, not fitting in, and not belonging. I've learned to mask my feelings well, mostly with humor, most people never know what's going on just below the surface. It's a full time job just to keep a lid on it. I hope that I'm able to keep control of my feelings, somewhere between sanity and insanity. I wouldn't like myself if I went too far over the edge and neither would anyone else!

The way this world is going today, I'm not sure that I want to fit in anyway. Whatever happened to taking responsibility for yourself and your own actions, honor, pride and country? JFK said, "Ask not what your country can do for you, but what you can do for your country". It seems like it's going the wrong way to me.

One emotion that I'm having a bit of a hard time putting into words is the way I feel watching these soldiers coming home from Afghanistan and Iraq. There is all this pomp and circumstance. You would think our fathers were coming home from WWII after being gone for years. Don't get me wrong, I harbor no ill feelings toward them. In fact I'm proud of them and God bless them. It's the way it should be, but I remember all too well what it was like for my friends and myself coming home from a war that most of us were proud to have been in. We did not expect crowds of flag waving people, but we didn't deserve what we received. Now, some forty years later, we get a welcome home parade. That's nice, but I say keep it. Too

81

little, too late. We almost had to hide our pride in being a Nam vet. Now they tell me that there are over 8 million guys that claim to have been there, so some of 5.5 million are liars. There's a bumper sticker that says, "I was a Vietnam Vet before it was popular".

Ski, Sgt. Hoch, Gary, Todd, and Jerry in Las Vegas

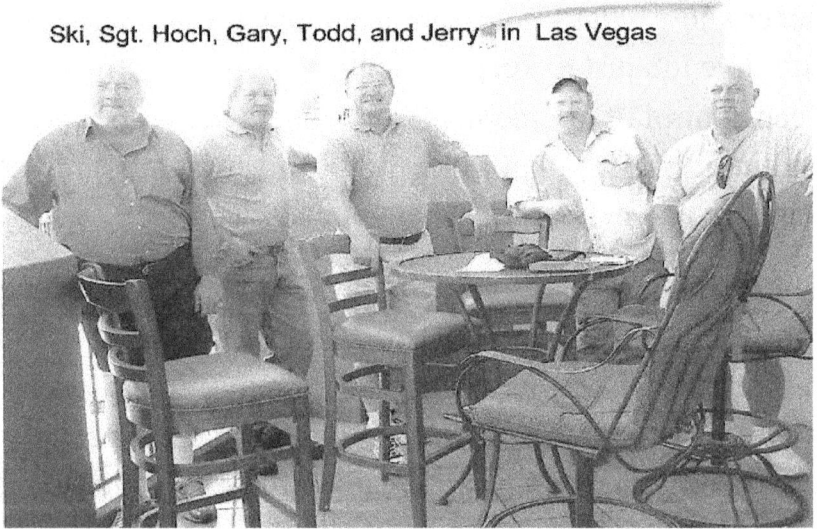

Our Old Tank Crew in Vietnam and Sgt Gary Mefford

Getting Together Again
Written by Todd Phillips

I spoke earlier about SGT Hoch putting me on his tank. The other two guys were Jerry Holly and Richard Gerszewski (Ski). Jerry was the only guy that I kept in contact with after the Nam. For some reason we just hit it off. He was my best friend.

I was working out in California 2004, when Jerry called me. He said "You're not going to believe this, but after years of trying, I finally got Sgt Hoch's phone number". I asked him if he was going to call him. He said, "No, You do it."

Well, I did. After 35 years we all got together in Las Vegas for the first time. It's hard to believe when you think about it, but I only spent a couple of months with Sgt Hoch 35 years earlier at which time I was 18 and he was 21 and I still picked him out in a crowd in Vegas. What're the odds? You can't get odds like that in Vegas. Clyde, Ski, Jerry, and I had a great time. I can't tell you how good it was to see them. It was a short time later we found out that Jerry had cancer. We met one more time in Washington, DC. Jerry and Linda, Clyde and Deb and myself. When we met in Vegas Jerry was about 235 lbs. When we met in DC he was maybe 135 lbs. Jerry passed away November 10, 2006, the Marine Corps birthday. It was a fitting day, for Jerry was a Marine through and through. I called Sgt Hoch and he flew up to Kenosha. We were both pallbearers for Jerry. Then not long after, we found out Ski had passed, so I guess Clyde and I are all that's left.

Jerry Holly, Clyde Hoch and Todd Phillips in Washington, DC, 2006. The last time we saw Jerry. Jerry received a Bronze Star and Purple Heart in Vietnam and two awards from the Great Lakes Police Department for saving people's lives.

Back row: Gary Mefford, Clyde Hoch, Lt. Peksens, Doc Forsyth, Ralph Schwartz. Front: Unknown, Fred Hoekstra in Washington, DC

Some of the guys have been meeting in DC for years. I started meeting them also. They met on November 10th and 11th. The 10th being the Marine Corps birthday and to pay respect for Jerry, the 11th being Veterans Day. Now one of the guys who went to DC has passed. In four years we've lost three of the gang. Is this part of getting older or because we are Nam vets? Good old Agent Orange! To anyone who is a Vietnam Vet, no matter what branch you served in, you owe it to yourself to go to the wall and DC over those two days. There is a brotherhood there that you cannot find anywhere else.

I was lucky enough to get ahold of a friend about three or four years ago that I was in tank school with in 1968. We were in boot camp at PI at the same time, but we never met until good old "Ocean Side" tank school at Delmar. His name is Jan Wedling. Now let me tell you about this guy and his family. In his family there are four boys and one girl. I don't think that his poor mother has had a good nights sleep since about 1965. There is John, Joe, Jim and Jan. all four of them went into the USMC, all four went to the Nam (and returned). All four became police officers in Mansfield, OH. The youngest child in the family is a girl. Her name is Mitzie. Not only is she absolutely beautiful, but she was a cheerleader in high school and a model. I wonder if she ever had a date in high school with four Marines, Nam vets, and police officer brothers. Poor girl. I bet if she did, the poor guy was probably shaking in his boots. I was lucky enough to write to Mitzie when I was in Nam. I drove up to see her in 1970. We should all be so lucky to have family like his. They are a very close and loving family.

ME
Written by Todd Phillips

Out of 58,479 plus names on the wall I'm sure that at least one of them would have made a bigger and better difference than I did. I think the measure of a man's life can be measured not by his wealth, but rather by his family and friends. All I ever wanted out of life was a happy loving family. I missed it as a kid, tried it twice in marriage and failed miserably both times. Maybe I should have paid more attention to June and Ward and Ozzie and Harriet. I've worked all over the country, never lived in one place more than 5 years and drank way too much for way too long. Life hasn't been all that bad. Family - not so good, but I'm lucky enough to have three or four good friends.

The one thing in my life that I'm proud of is being a United States Marine and a Vietnam Vet. That's a brotherhood that no one can ever take away from me. I've had people tell me that they were not affected by their service in Vietnam. To them I say bullshit, what you are today is because of everything you go through in your life, including Vietnam.

I belong to the VFW, American Legion, the Vietnam Veterans of America, and the Masonic Lodge of free and accepted Masons in Ohio and the USMC Vietnam Tankers Association. Out of all of the organizations I belong to, the USMCVTA is one of the best. Through them and going to DC I've been able to make contact and keep in touch with some of the men that I've served with. Including Sgt

Clyde Hoch-TC, Dave (Doc) Forsyth-corpsman, Gary Mefford- communications chief, Lt Peksens-XO (executive officer), Ralph Schwartz-TC, Doug Srivner-Flamer (flame tank) Jim Littman-tanker, Frank Carr-tanker, Jan Wendling-TC, John Wear-president of the USMCVTA, Dick Carey-founder and past president of USMCVTA. Deceased: Jerry Holly- best friend, Fred Hoekstra – mechanic, Richard Gerszeski-tanker. A few guys I'd like to find are Ronny Lyons, Ray Jones, "Chief" George P. Pumkinseed, George Watts, Sgt Coin and Ike England.

On a final thought, God bless all of you Nam vets out there.
We spend late spring and early summer of our life with our thoughts being dominated before, during and after Vietnam. Some of you have done better than others in the balance of summer and fall with getting along with your life. Now that winter is coming on, there are not many of us left. I hope that you can leave the war behind you and live the rest of your life in peace.
S/f Brothers
Cpl Todd Phillips 2489739
C company 1st Tank Battalion, 1st Marine Division '68, '69

M-48A3's at Battalion in Danang

Charlie Company headquarters near Marble Mountain

Marshall (Jamie) Jameson
Written by Marshall Jameson

After we torched the hooches and started across the clearing, during that unknown operation, we came into a grave yard. The TC (name slips me) instructed me to drive through the graves avoiding as many as I could to lessen the risk of throwing a track. (Vietnamese buried their dead sitting up and piling dirt around them, which made mounds). He also told me to drop the driver's seat and use the scopes to view out. I saw movements of straw hats bobbing up and down to our left as we started into the grave yard. I informed the TC of the movement, to which he replied "likely farmers".

 In a few minutes we were in deep shit. I was hearing small arms fire from a distance when the compartment became full with thick black smoke. In a matter of seconds I realized my comm. cord (cord that attaches from the tank helmet to the radio) was severed, leaving me with no way to communicate with anyone. Over the obvious confusion I could hear Swartz (the gunner) yelling over all that engine noise "I'm hit, I've been hit in the leg." I could hear the voice of the TC yelling at him to traverse the turret. Swartz kept yelling he was hit.

I began passing out and coming too and passing out again. I still did not realize I was hit, but I realized we were sitting ducks. The tank wasn't moving; the gunner and the TC were not paying any attention to the situation. I did the only thing in my power to do. I put her in reverse and stomped the accelerator; cutting the steering wheel to a hard right. The next thing I remember I was on somebody's back, crossing a field of brown, knee high grass. I could see a chopper in the distance. I remember the door gunner leaning over me as we lifted off and asking me if I was going to make it; to which I replied "Yeah!"

I do not remember, not one round of any kind leaving the tank. Not one shot. I do believe had I not taken action, we (the whole crew) would have died that day. I don't say that with any pride, just self satisfaction that I did the right thing.

Written by Clyde Hoch

I was with Marshall Jameson on this operation. It was a two tank operation or light section. Jamie was driver on one tank and I was tank commander of the other. We were attached to a platoon of Korean Marines. The Republic of

Korean Marines did not have tanks in Vietnam so we were attached to them to support them.

This incident took place in 1968. His tank took two hits from an RPG. I heard the explosion and saw dirt fly low and from the opposite side of his tank. I wrongly assumed it was a mine. One RPG ricocheted off the side of the turret, this was the one that caused shrapnel to hit Jamie. The other penetrated low on the hull.

The last time I saw Jamie was on the back of a Korean Corpsman on his way to a MedEvac chopper. For many years I wondered if he lived or not. He was the most well liked person I ever met in the USMC.

For more information on this incident see the book "Tracks Memoirs of a Vietnam Veteran" by Clyde Hoch

M-67 Flame Tank Firing Its Main Gun
Photo supplied by Marshall Jamison

Gary Young
Written by Clyde Hoch

Terry Hunter   Dick Russell   Gary Young

Four of us came right out of boot camp and infantry
training to Second Tank Battalion, Camp Lejeune, North
Carolina. We were all in the same platoon, all bottom of
the barrel privates. Donald C. May, Robert Lee Alexander,
Gary Young and myself. May was a good guy from
Buffalo, New York. He did well at whatever he did.
Alexander was a wild man from Ohio who later located to
California. I say wild man because he was always getting
us in trouble. He had a knack for making it all seem funny.
Because of his humor, he got away with almost everything
he did. Young was a nice quite country boy, raised on a
farm in Kentucky. He probably had the most common
sense of all of us. I remember asking him what time it was,
he would look at the sun and be very close to the right
time. I remember people would say "Young where you

from?" His reply was always "Tucky!" I remember his light hair and the grin that he wore most of the time.

I remember one of us asking, "Where are we going to tank school?"

"We will train you," was the answer. We learned tanks from our Sergeants. I guess they did a good job. Years later someone asked me where I went to tank school. I said "I never went to tank school".

He said, "Don't that figure, you taught me so much."

We were told we were going on a Med Cruise. The U.S. has a fleet of ships and a battalion of Marines on the Mediterranean Sea at all times to protect American interests. We were in the Marines for probably six to nine months by now. We loaded our tanks on Mike boats (landing craft mechanized). One tank on each of the two Mike boats and three tanks on a U boat (landing craft Utility). A tank platoon is made up of five tanks. The first three tanks are the heavy section and the last two tanks are the light section. The boats drove into the flooded back of the ship. They raised a huge tail gate on the back of the ship, pumped the water out and away we went for the Mediterranean.

The USS Casa Grande, LSD 13 (landing ship dock)

The ship we were on was the USS Casa Grande, LSD 13. It was our home for six months. We lived in a small compartment at the bottom of the ship. The canvas racks were close together – so close that when you rolled over, you rubbed the guy above you – and piled eight high. Everyone wanted the higher racks. If people became sea sick or sick from over – drinking you were better protected in the higher rather than the lower racks. When you climbed to the upper racks you almost always stepped on someone, somehow.

Since the four of us were the same rank, age and experience level, we hung out together most of the time. We had lots of fun together. Imagine you are eighteen years old in a foreign port where hardly anyone knows you, all the alcohol and good looking women you could handle. What a great life!

After the Med cruise we spent another six months together. Then Alexander and I were told we were going on another Med cruise. Wow! Party time again and this time we knew what was going on. This was going to be great! I have no idea what happened to May. I never heard of him again. Young was going to the Nam. We knew we would all wind up there eventually. We knew we had the same fate.

As you can expect, Alexander and I had a great time. When Alexander got us in trouble I learned to keep my mouth shut and let him do the talking. Alexander came to visit when we were both in the Nam. He was in "B" company and I was in "C" company. He was a great guy and I wish I could have met up with him after our service years.

I learned that Young was killed in the Nam. An RPG (rocket propelled grenade) penetrated the tank turret and struck a white phosphorus round inside the turret and the round detonated. I heard many rumors but little of the real details.

Many years later I had the privilege of talking to his brother Lynn Young. Lynn told me I should talk to a Louis Ryle. He was the driver of Gary Young's tank when it was hit. I called Louis who has a welding business in Lovelock, Nevada. After finally getting ahold of him, he gave me the details.

Corporal Gary Young's tank B-25 (B signifies B company, the two signifies the second platoon and the five signifies the last or fifth tank in the platoon) was struck by an RPG on September 10th. 1967 at 16:55 hours. Gary was acting loader at the time. The gunner was killed instantly, his name was James Wilson. The driver Louis Ryle was wounded, his helmet blown off and he had shrapnel in his back. He suffered hearing loss for some time.

Ryle pulled Young and the tank commander (Gunnery Sergeant Harold Tatum) out of the tank, with the help of some grunts. The grunts were worried about the rest of the rounds detonating close to them as the tank was still smoldering. Ryle climbed back into the burning tank and moved it to safety. He backed the tank down the hill off the berm near a flame tank. As Ryle was getting out of the tank an RPG struck the flame tank.

Ryle said there was a lot of confusion, as is always the case in combat. He could not communicate with his crew as his helmet was blown off and he didn't know where it was. He was in the drivers compartment and wasn't sure what was happening inside the turret. To make matters worse, the enemy had USMC flak jackets, helmets and uniforms.

Ryle stayed with Gary Young and Gunnery Sergeant Tatum all through the night. The fighting was too intense to MedEvac them out during the night. He said they were hurting. There is not a tighter group of people than those who serve in combat together. He felt they would both die the next day. White phosphorus is a horrible thing to have on you. It will burn under any conditions, under water or mud. The only way to get it off is to cut it off.

Ten days after this incident Gary Young succumbed to his wounds. The next day Gunnery Sergeant Tatum died of his wounds. They were both severely burned from the white phosphorus.

All of the crew received Purple Hearts. Ryle was the only survivor of the tank crew. He received no decorations for his part in saving Gary Young and Gunnery Sergeant Tatum or moving the still smoldering tank away from the grunts. Gunnery Sergeant Tatum was awarded a Silver Star many years afterwards for this incident.

Soon afterward, Lynn Young, (Gary Young's brother) arrived in the Nam. He was put in "B" company, second platoon, tank number five or B-25 the tank that replaced the destroyed tank that his brother Gary was mortally wounded on. Gary was truly a great American Hero. I will never forget him.

Gary Young's Grave Marker

Lieutenant Richard Peksens
Written by Lieutenant Peksens

Platoon Commander Lieutenant Richard Peskiness
and his Platoon Sergeant

The following stories have appeared in the "Sponson Box"
the magazine of the USMC Vietnam Tankers Association.

I was born in Boston and after college entered OCS (Officer
Candidate School). I spent eight weeks at Quantico in 1967
and graduated from TBS (6 months) in May 1968. After
attending tank school at Pendleton, I arrived in Vietnam and
was assigned to 1st Tank Battalion and subsequently, to take
over the platoon of Lt Parrish who had just returned from the
Cua Viet with 2/26 and the BLT (Battalion Landing Team). I
recall multiple nighttime adventures around the rocket belt
where I would usually take the place of either Sgt. Hoch or
Sgt. Coco. A few memories of that time include the sapper
attack in August on the bridge below Tk Bn, the time the
ARVN left their hill to the right of our position and came
streaming through our wires to our outpost (we almost
opened fire!). The night we sank the two canoes crossing the
river below our position, firing on NVA (North Vietnamese
98

Army) officer (after pretending not to hear the NO FIRE from our CDC), and listening to Hanoi Hanna at night on our tank radios. Also, my first time in getting to use a Starelight scope and firing a LAW….good stuff. I remember one day setting up along 11th Marine perimeter (artillery) when I jumped on the slope plate, and my loader mistakenly fired the 30-cal between my legs….where my chest had JUST BEEN!

We were off to 2/26's position and the adventures in the Riviera and monsoon floods. I remember my next to last day with the ROKS (republic of Korean Marines) when we were fired upon by a Marine patrol from 2/1 and we returned a HE (high explosive) round that landed near our CP Cpl Mefford (Communications Chief). He said he couldn't take our SitRep (situation report) because they were receiving incoming - our round! We never confessed to this overshot. During this encounter, I had a round bounce off my hatch while conversing with a Korean Major.

After leaving Charlie Company, I became the OIC (officer in charge) of an ontos unit atop Three Fingers… stories about this assignment including the death of my corpsman from a mine blowing up his jeep on his way to Tk Bn.

Then on to become XO's (executive officer) with Bravo Company. After a short time on Hill 55, I was sent to oversee two tank platoons at An Hoa. This was a place of daily road mines and incoming artillery and rockets. We lost one TC (tank commander) from a bouncing betty along the road from An Hoa to Liberty Bridge and another TC who fell off the front of his tank while moving.

We had three other tankers killed at An Hoa during my stint....two from incoming and one in the Arizona. We had 5 who also lost limbs from the daily incoming....not a great place to stay! I had a tank go to Bn, where two people were wounded when the ammo dump blew in Danang. The ammo dump also was blown at An Hoa...big fireworks. We had multiple sapper attacks at AN Hoa where we got to do some "turkey shoots" as the NVA tried to get back through the wire.

Following this, I returned to Charlie Company as XO and personally got a few trips along the MSR with the famous RAT Patrol (tank and a jeep with a 50 cal)...not a good idea as can be seen from the next photo.

From the 2/1 position, I supervised the move to LZ Baldy and Ross and, after a 3 month extension, got a 6-By ride back to Danang the night before my Freedom Bird flight. As, I was "acting CO" just before my departure, nobody ever took care of my Service Record which was kept at BN. As such, although I spent my entire tour in the field, there is no record of my ever having participated in any operations... Thanks for that!

I got back to Lejeune and became the S3 for Ontos Battalion 2nd Marine Division before being discharged.

Piss-tube Encounter
Written by LT. Peskiness

I arrived in Nam in mid-1968 as a new brown-bar (Second Lieutenant) with a checkered record from my wild days at Quantico where I fought tooth-and-nail against conformity with less than beneficial outcomes. I was one of three 1800 (tracked vehicle) officers to arrive on the commercial flight into Danang after spending three days and nights visiting all the attractions that Camp Hanson and Okinawa had to offer. In other words, I was exhausted and hung-over.

We spent the first night in officer quarters adjacent to Danang Exchange where we received the "Rules of Engagement" speech which essentially allowed the VC/NVA (Vietcong/ North Vietnamese Army) to operate along the borders of adjacent TAOR's (Tactical Area of Responsibility) without fear of retribution due to the complexity of trying to coordinate fire missions between multiple entities including the ARVN (Army of the Republic of Vietnam).

The following morning, we were given our assignments. The other two 1800 officers were assigned to the 3rd Marine Division. One was given a 0300 grunt assignment and the other was sent to 3rd tanks. I was assigned just around the corner to 1st Tank Battalion which was a quick jeep ride through Dog Patch and then south along the Danang runways and old Route 1 until we turned right at the French Bunker and climbed the hill to Battalion located directly north of the 8th Marines Artillery Battalion.

101

Upon arriving, I was told that I would be assuming command of the 3rd Platoon of Charlie Company which had just returned from the BLT (Battalion Landing Team) at the Cua Viet River where LT Parrish, the current platoon leader, had received the Silver Star. I was told that I would need to wait until the next day to receive a weapon and other combat necessities. I was then taken to a spot on the berm where I was expected to muster in the event of a rocket or sapper assault. I was also told to sleep with my boots on. Naturally, I was a bit nervous about facing drug-crazed sappers without a weapon. I had seen what happened to John Basilone at Guadacanal and Sergeant Striker on Iwo Jima! That night I settled into my new bunk fully dressed awaiting imminent attack. At about 1:00 AM, the siren sounded and everybody began jumping out of their cots and disappearing into the murky night. Without anyone to direct my exit, I left the tent under the illumination of the high-flying flares and frantically tried to get my bearings to my assigned spot on the berm. I could hear the explosions of the incoming rounds (122 millimeter rockets) and outgoing artillery from the 8th Marines.

By the time I had zig-zagged to the berm, the COC bunker had already sounded the all-clear. I was now forced into the daunting task of finding my way back to my bunk in utter darkness. My jumbled nerves had caused my bladder to overfill and I immediately started to search for a "pisstube". I had previously "snapped-into" the pisstube while in OCS (officer's candidate school) and staging at Pendleton.

I knew the standard issue pisstube to be an approximately 2 inch diameter pipe extending from a ground sump and topped by a large can with a mesh cover. A Marine simply urinated into the mesh, apparently the mesh prevented the passage of calcified stones into the sump, where the urine then leached into the ground after passing through a layer of lye which broke down the toxins. After stumbling through the dark, I spotted the pisstube which appeared to be enclosed in a wooden 3-sided shed with a roof, which was useful when emptying the bladder during the monsoon conditions. As I approached with ever-increasing urgency, I noted that a circular cement platform had been constructed adjacent to the pisstube to obviously keep a senior officer's boots dry while performing their restroom duties... these officers had thought of everything!

As I stepped onto the platform, I learned my first combat lesson to always expect the unexpected as I plummeted into a 55-gallon drum of lye and urine. What appeared to be a cement platform was actually a mesh screen and the "pisstube" was actually an aeration device.

With my platoon introduction the next morning, I was now submerged to my waist in urine and toxic chemicals. After removing myself from the barrel, I was pleased to see that my accident lacked witnesses as everybody returned to their respective hooches following the alert. I luckily found some outside shower where I was able to wash my quick-drying utilities and attempted to clean my leather boots.

The next morning I avoided breakfast, due to the foul odor emanating from my boots, and was introduced to my platoon wearing wet boots that still smelled of urine and lye. Luckily, the quartermaster provided me with jungle boots in the afternoon and I was able to "deep-six" the incriminating evidence. I had suffered from athlete's foot fungus since playing HS football. Miraculously, the upside of my encounter with the pisstube had cured this long-standing ailment apparently due to the daylong immersion in urine and lye.

After leaving the Marine Corps, I later learned that soaking your feet in urine was a traditional off-brand cure for foot fungus... I recommend that readers try this at home.

Later in my tour, I would also have additional excrement adventures while burning the crappers. I would learn to be careful emptying the residue and avoiding inhaling the black smoke. Once while visiting battalion, I was sitting on the 3-holer reading a magazine (Leatherneck?) when the Colonel entered. I was torn between choices of proper etiquette. Do I stand and salute? Do I jump up and essentially moon the CO (commanding officer) while trying to leave or do I remain and chitchat with the CO and wait until he has finished his business? I chose a quick salutation and exit. These matters are never satisfactorily covered in OCS during the "WHAT NOW LIEUTANANT" series.

The conclusion of this story is that although I didn't intentionally wet my pants during my first combat encounter, the result was the same. It has taken me 30 years to get this confession off my chest.

Vacationing in the Riviera Circa 1969

Written by Lt. Peksens

Marble Mountain - the south by Hoi An, the west by Route 1, and the east by the beautiful South China Sea. The Riviera was the most heavily booby-trapped AO (area of operation) in I -Corps.

Charlie Company 1st Tk BN (tank battalion) was headquartered a few miles south of Marble Mountain, at the end of the MSR (military supply road) within the CP (command post) of 2/26 (later 2/1). Our tanks deployed within the TAOR's (tactical area of responsibility) of the ROK Marines and the 1st and 7th Marine Regiments. With the exception of the South China Sea, this zone was nothing like the fabled French Riviera. As a platoon leader in mid to late 1969, a few interesting incidents remain in my mind. The facts are true, but the exact chronology escapes my rapidly deteriorating memory.

On one occasion, we had taken three tanks to accompany a grunt platoon sweep north of our CP. After uncovering and demolishing a number of bunkers, we set up in a position close to the MSR to find land mines nearby. Near midnight, incoming rounds started exploding and moving toward us. The rounds were bursting in the air and moving straight at us! Thinking we were under mortar attack, we manned the tank's weapons as the rounds began exploding directly overhead, raining down pieces of hot shrapnel over our tanks. As we frantically called on our battalion frequency to report the impending attack, we learned that we were the victims of H & I fire from a 11th Marines 105 millimeter battery in Danang. After the incoming ceased, we began crawling out of the turrets to assess the damage. Miraculously, only a few grunts had been wounded as most of the troops had wisely sought refuge on the undersides of our M-48's. Had our tanks not been there, I'm sure the grunts would have taken at least 50% casualties as the air-bursts were directly overhead for two or three minutes, which seemed like an eternity.

On another occasion, we were crossing a small tributary just south of Marble Mountain next to a CAP village called Nui Kim San. We knew a number of amtraks had encountered mines at this crossing point because it was a common route for them. The soft soil near the water crossing made an ideal spot for planting explosives. Having seen too many John Wayne movies, I crossed the river with my K-bar (knife) and started probing for buried mines. Satisfied that my EOD training had not been in vain, I called my tank across with Corporal Aikens at the wheel. About one month earlier, Corporal Aikens had been blown from his driver's seat along the MSR when my tank hit a mine while traveling to our nighttime setup position at a small bridge next to Route 1.

Just as the tank started out of the water, it hit a large mine which blew me about 20 feet from the slope plate. Corporal Aikens body landed 10 feet closer to the tank. My helmet had ended up in the river and my flack jacket was chock full of large fragments of shrapnel. I had been sand blasted and punctured an eardrum, but only had a few small pieces of shrapnel in my one forearm. The sprocket and a road wheel had blown off the left side of the tank along with a section of the track. I looked up and was surrounded by Marines from both our group and the local CAP (combined action platoon) unit. They couldn't believe that my driver and I had survived the explosion essentially unscathed. After dusting myself off, we called the retriever and later were towed to our CP. I had our corpsman remove the few pieces of shrapnel and went to 1st Med the next day to see about my blown ear drum. They gave me an ear plug and ear drops (no purple heart) and told me to avoid loud noises for a few weeks until the membrane repaired. I believe that I took that opportunity to take R&R in Australia where I could avoid most loud noises.

"C" Company headquarters near Marble Mountain

(Left to right) Lt. Dan Guenther, platoon leader of an Amtrack platoon,
Sean Flynn (son of Earl Flynn) and unknown.
This photo was given to Lt Peksen by Lt. Guenther.

Sean was an actor and a singer. He became a guide for big game hunters in Africa. He also tried being a game warden. Later he became a photo journalist where he traveled to Vietnam to photograph the war. In 1970 he and a fellow photojournalist were traveling in Cambodia by motorcycle, when they both disappeared. He was never heard of again. It is believed he was captured and held prisoner and later killed by the Khmer Rouge of Cambodia or the North Vietnamese Army.

The photo above was taken while Sean was visiting Republic of Korean Marines in Vietnam.

While leaving our CP for a day operation in the Riviera, our three tanks stopped by a river where a company of Marines from 2/1 were stationed. Suddenly there was a loud explosion from within the company CP and I ran to see what had happened. Somebody, perhaps attempting to relieve themselves, had tripped a large booby-trap within the company compound. I soon found learned that it was a LT Puller, Chesty Puller's son, who had graduated a few months behind me from OCS.

He had lost portions of three limbs and was a bloody mess. A number of Marines were applying tourniquets to his limbs and trying to get him onto a stretcher. He kept saying, "Don't tell my Dad," as they transported him to the MedEvac site. I was sure he wouldn't make the helicopter flight to 1st Med. Many years later, while living in Norfolk, Virginia, I was amazed to find out he not only survived but had successfully run for state legislature.

Following this incident, our tanks prepared to cross the river towards the South China Sea. Corporal Aikens walked across the 30 foot span and found that the water was no deeper than 4 feet which we could traverse without additional fording gear. We started crossing with a grunt in front of each track, but when we got to midstream, we started lieaning to the port and sinking! Apparently, the left track had hit a large bomb crater and we were rapidly sinking into the silt of the river bed. I looked down and observed water at the level of our breach while my gunner and loader had already taken refuge outside the turret. As the tank continued to sink, I was worried about Corporal Aikens who was now at least 4 feet below the surface.

After what seemed an eternity, he finally surfaced for air. I asked why he was down for so long and he stated that, while holding his breath, he was trying to put the treads in reverse. Who says Marines can't multitask? At any rate, the twin diesels "died" with only the turret and the surface mounted 50 caliber above the water line. We then got another tank and ran two tank cables to the submerged tank and tried to pull it to free of the salt-laden river bottom. After this failed, we used both tanks in a conga line…this also failed.

We then called back to Charlie Company and asked for the retriever and were told that they would be escorted out by two other tanks and would arrive at about dusk. Looking across the river, we noted a small number of NVA on an elevated sand dune about half a click away were observing our predicament. We didn't want to fire on them for fear that they could be a significant nuisance to our tank retrieval, but it gave us good reason to hustle with the extraction.

When the retriever arrived, he hooked onto the two tanks which were already connected and slowly pulled the submerged tank from its watery grave. With the line of multiple vehicles, we started back for the MSR without the ability to make any turns. This proved interesting when we found ourselves heading for an occupied village where a number of papasans were wilding waving us away. It was growing dark and we had no time to banter with the locals. We proceeded to run our daisy chain directly through the village and over a number of grass and plywood hooches….so much for winning the hearts and minds of the people! I believe that I may still hold the record for sinking a tank in Vietnam.

A 1st Marine company near the leper colony, in the Riviera, had been experiencing extreme enemy probes by sniper fire for a number of days and actually had NVA crawling into their fighting holes. The grunts had asked if tanks could be sent to protect their tenuous position. Subsequently, we left our CP with three tanks and grunts aboard to reinforce the Marines under siege.

We needed to make a stop along the beach when a Marine with active diarrhea had fallen off the back of a tank while trying to relieve himself which resulted in a compound leg fracture.

When we arrived at the company CP, we immediately noted that our defilade position was less than ideal. On the west side of our slightly elevated position was thick underbrush which approached to within ten meters of our tanks. Also there was no defilade for either the driver or the slope plate. We were positioned with one tank pointing south, another north and one west.

During the day, we noted NVA running across the open across road leading to the beach. They were traveling in small groups which gave the grunts no time to fire before they scampered into pine copses 200 meters from our positions. As soon as it began getting dark, we started getting blooper rounds within our perimeter. We responded with our own bloopers and grenades, but failed to silence the harassment. Next, we started getting RPG fire aimed mainly at the tank bordering the south portion of the camp which was the most exposed. The TC, Sgt Hoch, came to me and asked if we could pull back as we had seen

111

multiple hits in close proximity to his tank. Unfortunately, pulling back would still leave him exposed. The NVA were also getting onto our frequencies and we were forced to change more than once. The more the night extended, the closer we felt to being overrun. We dug holes under the back of our tanks and were keeping two people safe while two manned the guns. At one time Sgt Hoch's crew had come to my tank to check the frequencies and had received some blooper shrapnel as a reward. They simply pulled out the small pieces and went about their business. I finally came up with a possible solution for spoiling the NVA plans.

Each tank would plan on firing three "shotgun" rounds simultaneously encircling our area by rotating the guns between shots. I checked with the grunt lieutenant and he agreed to the plan and arranged for his men to move back parallel to the tanks at the appointed time. At the given hour, we let loose with the shotgun rounds.

Trees shrubs, and our own concertina went flying through the air and we received only sporadic return fire which lasted only minutes. The next morning, we found numerous blood trails within feet of our concertina wire. An RPG had been abandoned a few feet outside the wire in front of Sgt Hoch's tank and a dead water buffalo was found about 100 meters away with a NVA heavy MG trapped underneath the carcass. There is no doubt in my mind that an attack had been imminent. Because of the thick cover, we were forced to leave the next day after helping to clear good lines of fire by steamrolling trees and shrubs with our tanks along with a blade tank which had been escorted to our area. On the way back to the beach, we passed directly through the Leper Colony and were amazed by the view of children and adults living in grass huts, missing limbs, and seemingly without medical help. We, of course, were happy to still be in a single piece.

I would leave Charlie Company to become an Onto's platoon leader at an outpost called Three Fingers along the route from Danang to Hill 10 and hill 55. I followed that assignment by assuming the position of "second" Bravo Company XO at An Hoa. Bravo Company had a "real" XO on Hill 55 with the 7th Marines, but the distance to our platoons at An Hoa required a person to interface with the Command Bunker of the 5th Marines. Later, during an extension, I would return to Charlie Company as XO during our move to LZ Ross and Baldy, but they are stories for another time...

Dick Peksens
1st Lt USMCR

Corporal Jan (Turtle) Wendling

Written by Jan Wending

In July of 1969, Alpha Company, Third Tank Battalion, Third Marine Division was located just south of the Demilitarized Zone (D.M.Z.) in South Viet Nam. Third Platoon was stationed at Con Thien, (Hill of Angles) two thousand meters south of The Ben Hai River, which ran through the center of the D.M.Z. at the 17th parallel, first and second platoons were stationed three miles south of Con Thien on a Fire Support Base (FSB) called C-2. H&S (Headquarters and Supply) were located in our rear at Quang Tri. We also had flame tanks (M67A2) that spewed Napalm out of their main guns. The "Flamers" had a three man crew and were affectionately known as Crispy Critter and The Devils Disciple.

Each platoon consisted of five tanks with four man crews - the driver, loader, gunner and tank commander. Our tanks were Patton M48A3's which were considered "medium" gun tanks in the Marine Corps but was the main battle tank. The weight of the tank was 54 tons with a combat load. Each tank had a 90 MM main gun with a coaxially mounted .30 cal. machine gun. The commander's cupola had a .50 cal. machine gun in it. There was a M3A1 sub machine gun (grease gun) located on a rack on the inside turret wall on the right side of the turret. Each crew member was issued a 1911 .45 cal. Pistol also. Eight grenades were also carried. Personal weapons were carried by each crew member also such as M79 grenade launchers, M14 and M16 rifles, shotguns, Law Rockets and even cross bows.

Our Company Commander was Captain Wunch, a 25 year old tall blond, medium built man from Feasterville, Pa. He came to our company after serving in an intelligence unit. He was well liked and respected by everyone in the Company. He was also a friend and mentor to me personally.

Our job at this time was running security for convoys coming up highway 681 from RT# 9 at Cam Lo Village to the south and standing Slot Patrol (off the side of hwy 681) where we sat in cutout slots just wide enough for our tanks with a dirt wall on 3 sides so only the turret could be seen. These slots were located off of each side of the road and spaced about 100 yards apart. We had a few grunts at each slot. We stood slot patrol all day long. We also ran sweeps through "Leatherneck Square". Leatherneck Square was an area which ran from Dong Ha north up RT# 1 to Gio Linh (FSB - Fire Support Base) just south of the DMZ, west to Con Thien, south to Cam Lo and then back to Dong Ha on RT# 9. Highway 681 was nothing but a dirt road that ran north out of Cam Lo village for ten miles and ended at Con Thien Combat Base. The road was very narrow and in some places dropped 10 feet on both sides into rice paddies. The area north of Cam Lo village for ten miles up to the DMZ was a "free fire zone" where you could shoot anyone in the area.

On July 25th we were advised by Capt. Wunch to get our tanks and crew ready because we were going to start an Operation going north and west of Con Thien. The Operation was going to be called Operation Idaho Canyon. These orders took everybody by surprise as it was already 1700 and we usually started everything we did at first light. We were leaving the safety of our base and going into the unknown. We left C 2 at around 1800 hrs. and headed north to hook up with 3rd Platoon at Con Thien. There was an abandoned Marine FSB known as the "Washout" half way up to Con Thien and we had to ford the creek there. There were two burned out tanks laying in the bush just west of the Washout that were blown up and destroyed in some long forgotten battle. It was said that they had to be napalmed for two days after the battle to weld everything together so the NVA could not use anything off of them.

Then we headed to Con Thien. The road was filled with the smell of diesel fuel and you could not see in front of you for all the dirt and dust the tanks were kicking up. Navy Seabee's dozed both sides of HWY 681 a couple of months earlier from C 2 to Con Thien for 1000 meters to stop ambushes. We pulled into Con Thien and hooked up with third Platoon. The grunt Company Kilo 3/3 was going with us. We were advised that one of the third platoon tanks had a road wheel locked up and another one could not traverse it's turret but they were also going on this Operation. A feeling of foreboding blanketed everyone and a stillness seemed to come over the base as everyone contemplated meeting the NVA in their own back yard. We headed west out of the base and had gone only a short distance when there was a problem at the front of the column. A 12 (Sidewinder) was mired in a rice paddy and A 13 (The Hulk) was also stuck. We took the tow cables off the slope plate in front of the driver and hooked up to A 13. We pulled them out and we hooked up to A 12, who was deeper in the mud, but finally got them out too. We then started heading north where we picked up an old French road and followed it up close to the Ben Hai river, where we set up a perimeter for the night. We pulled the tanks into a circle, like the old wagon trains and stopped for the night. By this time it was already dark. It was good to stop. We all looked like we were wallowing around in a pig sty after getting the tanks out of the paddy. Now as we set up a night defensive position it started to get scary. The grunts sent out their listening posts (LP'S). We were put on a 50% watch, which meant two crew members had to sit watch for half the night while the other crew members

117

slept and then we traded half way through the night. Being a tanker was not easy but those poor grunts had to go out on those LP'S every night. I have three brothers who were all grunts and they all spent a year in country. The stories they told would curl your hair. Grunts have big BALLS.

Night time in enemy territory is enough to make anyones hair go white, I'd been in fire fights (combat) before but it was usually in daylight. Nighttime was a different animal. They say that we ruled the day and the NVA ruled the night, which is pretty much the way it was. Standing watch is always spooky. You wonder, first of all, if there is some little gook out there who has his slanted little eyes looking down the barrel of his AK 47 with his sights on your forehead. You just "know" you saw something move out of the corner of your eye and you hear sounds that you've never heard before, right out in front of you. It scares the shit out of you. You wonder how you are going to react if the shit does hit the fan. Are you going to be faster than he is? Are you going to be luckier? All kinds of demons and phantoms run through your mind. All you want to see is your next birthday and your parents and your little girl friend back home. Most of all, you pray. You can never be TOO close to God in a combat situation. Mostly, you pray for daylight. At any second you could see the flash of a Rocket Propelled Grenade (RPG) or the knop-knop-knop of an AK 47. Then as darkness released its grip, it starts to get light again, now you know your prayers are answered. Now it was time to start the second day of Operation Idaho Canyon.

We started to move west towards the mountains. We came upon miles and miles of uninterrupted fields that fell away from us in giant wide steps. We had the Ben Hai river to our north and the sprawling fields to the west which turned hillier and hillier the closer we came to the mountains, which were quite a ways from us. All the tanks were inline with the grunts moving out in front of us. There were bomb craters everywhere and we all had to maneuver around them as we approached. As we looked north, we could see the Ben Hai river and beyond into North Vietnam. You could see every possible shade of brown and green. It was truly one of the most beautiful sights these eyes have ever beheld. Off to the southwest, we could see a platoon of grunts coming off of a wooded hillside. They were led by a young Lieutenant named Oliver North. We knew him as Lt. Blue, as the phonetic for north on our maps was blue. He was well liked and respected by his men. He and Capt. Wunsch had gone to Annapolis together. All of these images are burned into my memory and I can still see them today by simply closing my eyes.

We were advised that a NVA radio frequency had been picked up and one of the Kit Carson scouts (a former NVA who gave up and was now working for us) advised that the NVA had spotted tanks and infantry and that they were going to hit us at night or ambush us. This news really put the "Pucker Factor" into play.

As we got closer to the mountains, the fields turned into rolling hills. Small at first then they grew higher and higher until the mountains shot straight up to the sky. We had an air observer flying overhead in a OV10 (Bronco) airplane. His call sign was Sky King. He advised us that he observed what he believed to be a Company of NVA in heavy camouflage from north to south out in front of us. He then said he lost them in the vegetation. We then headed for the highest hill we could find and set up a perimeter on top of it. Sky King could not locate the NVA again but he fired a White Phosphorous rocket at the last place he saw them. Anybody who had a set of binoculars on the hill was trying to find the NVA, but to no avail. The LP's left the hill at dusk and we were put on another 50% watch for the night. I stood the second watch with my driver, Bruce Lester (Foxy) Fox from Brooklyn NY. My loader and former tank commander Eddie Miers from Athens, Texas and my gunner Bill Norwood MA took the first watch. As it got darker and darker, we believed that we were about to get hammered by the NVA sometime in the night. Everybody was on edge and I don't believe that many people got any rest that night. We were all elated as it started getting light, knowing we had cheated death once more, when, off in the distance you could hear THUNK-THUNK-THUNK, one right after another. I wondered what the sound could be when I realized mortars were being fired at us. We started yelling "RAINDROPS", which meant mortars, and everybody out in the open scattered looking for some place to hide. The first round went off in the middle of the perimeter with a distinctive CRUMP. They must have had a circle of guys running around the mortar tube dropping a round in as fast as they could. The blasts were deafening. There were shards of steel, dirt, rock, branches and everything else flying around the perimeter. They dropped 40 rounds on us and it ended as fast as it started. Screams of "Corpsman Up" were sounded and I wondered how many of us were hit. Bill (Creech) Franker from Calamut City, IL had shrapnel in his ass and back and had to be MedEvaced (Medical Evacuation) out. LT. Ralson from second platoon had been hit in the back by shrapnel but was not serious enough to be MedEvaced. HMS 3 Michael (Doc) Colton was struck in the forehead with a piece of shrapnel also but decided to stay.

We headed south off the hill wondering just when and where the NVA were going to hit us. We got halted again when the lead tank came upon a mine field. None of the tanks hit any mines though. We were headed to Hill 70 to set up for our last night of the Operation. A Staff Sergeant from second platoon, SSGT Schley advised the Captain that he was on this hill in 1966 and they got their butts kicked. He thought there might be a training camp for the NVA in the area.

As we neared Hill 70, A 52, a dozer tank, broke the final drive and couldn't move. We were at the base of Hill 70 and we passed A 52 as they worked on it. Hill 70 was a good size hill with half of it being wooded and lower and the other half being in the open and higher. I advised Foxy to head for the wooded side because I didn't want the tank to be exposed on the opened side. As we set in, I had Foxy back A 14 (The Bleeding Eyes) up to the rim of a bomb crater, so nobody could jump up on the back of the tank (NVA wise).

Captain Wunch called me and advised me that he wanted me and my crew to go back down the hill and pull A 52 up the hill. I knew if we did this, I would lose the security of my present position and I didn't really want to go, the Captain told me to disregard as he was sending one of the new tanks down to get A 52. THANK YOU, LORD.

It was hotter than hell by the time we set up which was about 1400 hrs. There was a creek at the bottom of the hill and some of the guys were filling up their 5 gallon water cans in it. I grabbed ours and filled it up also. As I got back to the tank I secured the can on the side of the turret. We kept a sawed off .12 ga. shotgun in the Bissell rack on the back of the tank and I noticed that somebody had put it on top of the cupola. We never carried a round in the chamber. As I picked the shotgun up it went off with a loud bang. Eddie Miers was standing beside me and I saw him grab the side of his head. I thought, " I just blew Eddies ear off." It had actually missed Eddie but about deafened him. I think every officer on the hill was standing by my tank wondering who the dumb ass was who had the accidental discharge (AD). Captain Wunsch called me everything but L/CPL "Turtle" Wendling and told me to see him when we got back the next day. I felt like a piece of shit. I hated that I almost killed Eddie but most of all that the Captain was pissed at me. I never did figure out who loaded that shotgun. I think I felt worse because I had just taken over as tank commander (TC) from Eddie Miers because Eddie was getting ready to rotate back to the WORLD. This was the first time that I was the TC in the bush. I didn't think it was a very good start. Luckily for Eddie, his hearing came back.

It started getting dark again and the crew ate C rations on the tank. We all talked about how we knew we were going to get hit. I wanted to be sure we had everything ready when it happened. We put ammunition close at hand for our weapons. I had M 14 magazines, M 16 magazines, .45 cal. Magazines and .12 ga. Shotgun ammo on my copula. I had grenades with the pins straightened readily at hand. We had these tanks, and they were nothing more than a mobile ammo dump. We carried 62 rounds of 90 MM ammo, 500 rounds of .50 cal. Machinegun ammo, 5900 rounds of .30 cal. Machine gun ammo and 360 rounds of .45 cal. Ammo, plus 8 grenades and anything else you could carry. I also had pop-up flairs in the Bissell (gypsy) rack. Most of the flairs were illuminated but different colored flairs meant different things. Green meant that friendly troops were coming into the perimeter. Red flairs meant that the perimeter was breached by the enemy, and no one wanted to see those.

We had grunt machine gun teams setting up to the right of the tank, over the lip of the bomb crater. My best friend, Wayne (Tuna) Tunison from Sarasota Fl. had A 11 (The Lemon) sitting to my right and the broken dozer tank A 52 had set up on my left. Tuna and I discussed protecting each others tank if the shit hit the fan. Tuna and I went through Tank School at Camp Delmar, back in November and December of 1968 and had become close friends.

The LP's were sent out and they set up their trip flairs and Claymore mines once they were in position. It was deathly quiet. Too quiet. Then at around 2200 hrs several large explosions went off and woke everybody on the hill. The explosions came from the LP out in front of our tank. The radios started to come alive. The LP had movement just off of their hill and several trip flairs were popped. The grunts then set off several of the Claymore mines. The radio operator could hardly talk from fear. He advised that they had at least 10 very dead November Victor Charlies (NVA) outside their lines. He also said they were being probed by a large number of NVA. The LP was advised to report on any more movement but to stay put until advised. We all knew our shit was weak now.

It got real quiet again and there was not any report from the LP. Eddie and Bill tried to get some shuteye but that was next to impossible. Foxy and I had the second watch and we both laid down on the back of the tank, it was 0200 hrs. It seemed like we had just laid down when all hell broke loose. It was now around 0300. As I opened my eyes, all I could see was green tracers zipping over the back of the tank over our heads and I could hear the CRUMP of mortars hitting everywhere. My first thought was that I would never see my girlfriend again. I told Foxy to lay still until we could get into the tank. If we stood up we would have both been shot dead. I said "lets go" and we both made it into the turret. The problem was, we couldn't fire the main gun with Foxy in the turret because the breach came back 13 inches and would have broke his leg. I told him to get to the drivers compartment and that I would cover him. He jumped out of the loaders hatch as I stood in the cupola firing the M 16 on full automatic. I thought for sure, the way he was jumping around trying to get to the drivers compartment, that I was going to shoot him but he made it okay.

Eddie loaded the main gun with Flechette (beehive) round. It had 8000 little steel darts that looks like a nail with three fins on it and came out like a big shotgun. You could set it to open up (air burst) up to 2500 meters. The beehive was devastating to anyone on the receiving end. What a great round it was. My personal favorite. Bill fired and we kept loading them. It was sheer mayhem outside the tank. Bullets and shrapnel of all kinds were hitting the turret. I could smell all the freshly blown up dirt and the sappy smell of vegetation being torn apart and thrown everywhere outside the turret.

Bill started firing the .30 cal. machine gun and was walking them all over our front when all of a sudden it stopped. Eddie took the gun off it's mounts and checked to see what had gone wrong only to find that the extractor/

ejector had broken. We didn't have a spare. The 50 cal. machine gun had never worked because of the way they made us mount them in the cupola. They actually laid on their side, which jammed the machine gun belts. They were also electrically fired. So we had no machine guns. Only the main gun and our personal weapons.

SSGT Shley screamed over the air. An RPG had penetrated his tank and he had part of his arm blown off. When he fell, his mike was keyed on the side of his helmet. Those guys on the open side of the hill were getting slaughtered. The first RPG fired hit A 41, the Captain's tank. The Captain had been watching NVA, through a starlight scope, as they prepared for their assault. LT. North was standing beside him and turned to tell his men to get ready when the RPG hit the cupola. Captain Wunsch lost his life immediately. Lt. North was blown off of the Captain's tank and was wounded. One of LT. North's machine gunners, Randy Herrod, pulled him to safety.

When A 41 was hit, they said over the radio that A14 was hit. I saw my best friend Tuna, jump off his tank to come and check on me. I yelled to him that it was A 41 that got hit and not me. It didn't take him any time at all to get back in his tank.

A 23 got hit by several RPG's. One entered the turret on the right side and penetrated. HM3 Doc Colton and Steve Dowdell were both killed.

If you look in front of the A-23 you can see a plugged hole from an RPG

Corporal Frank (Tree) Remkiewicz. Wounded 7/28/69

A-24 was hit numerous times and had many penetrations. They were to the left of the Captain's tank and down the hill. Cpl. Frank (Tree) Remkiewicz the TC, was wounded by shrapnel. The driver, Cpl. G. E. (Granny) Nappier was sitting in the driver's seat when an RPG struck the turret behind him peppering his neck and shoulders with shrapnel. The loader, Dave (Animal) Turner was sleeping on the sponson (tool) box on the side of A 24 when the next thing he knew, he was blown to his feet and was standing on the back of the tank with all his clothes blown off. His leg was broken in 15 different places, his hip was busted and he lost part of his foot. When he went to get inside the tank, he was hit by a mortar. He got down inside the tank and found that one of the fins from the mortar entered his head and was sticking in his mouth. They had to abandon the tank and Tree helped Animal back up the hill. Cpl. J. B. Renfroe was sleeping on the side of his tank when an RPG hit the turret above him knocking him off the tank. He was shot in the leg and caught shrapnel from a Chinese Communist (ChiCom) grenade.

A mortar went off beside my tank and blew my CVC helmet around my head. I thought I was dead. When I got my helmet back around I looked at the grunt machine gun team climbing back out of the bomb crater where they had been blown by the blast of the mortar. We started a fire in the bushes out in front of my tank with our tracers. You could see the NVA running in front of the fire and it was just like a shooting gallery. We called on 12th Marines at C

2 for illumination from their 105 howitzers. When they opened up outside the perimeter it looked like daylight. The steel casings they were firing came crashing to the ground with a WHOOP, WHOOP, WHOOP. If you were hit by one, it would kill you. We also called on Spooky. Spooky was a DC 3 prop air plane with mini guns hanging out the side. It was capable of putting a bullet in every square inch of a football field in 10 seconds. When he came on station you could have heard a bug fart. It got real quiet until he left.

We were trying to get MedEvacs for our dead and wounded. Two Chinook twin rotor helicopters finally arrived. Bullets were going through them and you could hear the pop,pop,pop,pop. Those pilots had very large testicles. We couldn't believe we had so many casualties. The grunts alone lost 17 people. I don't remember when it was that I had to send Eddie and Bill off the tank to man the tanks that were decimated. Everything that happened that night is like a blur now. I do remember most of it. It's funny how two guys that were in the same battle both remember it differently. You only care what is happening right out in front of you. There is so much going on that you can't possibly see or hear it all. We were fortunate to have made it out alive.

As daylight started to open it's eyes on a new day and we could see around us, it was surreal. The ground around us looked like a WWII movie. There were lots of blood trails leading out into the bush. The NVA liked to carry their dead away just like we did. We did find one NVA body. He was run over by one of the tanks and he was still under the track the next morning.

We had to tow one of the tanks back to C 2. The others made it back on their own power. We came through an area in the hills that looked just like the moon. There were no bushes and the trees were stripped of any bark or leaves. There were no birds, no bugs and it was full of bomb craters. The dirt was like baby powder and it poofed up when you stepped in it. This area was full of Agent Orange and we had it all over us.

We had a memorial service for the dead at C 2 several days after the Operation and some of the wounded attended. Out of all the wounded, only Lt. Rolston remained at C 2. All the others went home. I still keep in touch with some of them. Tuna died in 2006 from complications of Agent Orange. All of us that were there are truly brothers.

Cpl. Jan (Turtle) Wendling 2431841

Jan Wendling (left) in front of "The Hulk" A-13 March 1, 1969

After serving in the Marine Corps, Jan became a police officer and served with his three brothers John, Joe and Jim who were also police officers in Mansfield, Ohio. His cousin Gordon was also a police officer. He became a hand gun instructor for the police department.

Tanks in Vietnam
Written by Sergeant James F. Johnson

My first tour was with 1st tank battalion, "C" company, December 1966 to June 6th 1967. My tour was cut short due to an injury. I was sent to Japan for surgery and eventually sent to a hospital in the U. S. We worked to the west and to the north of Danang. I had a couple of experiences on hill 55.

We hit a mine which tore up the track towards the middle of the tank. It blew off two sets of road wheels. I got out to fix the track and as I was under the tank with my leg against the road wheel arm (which had lost its road wheels) some idiot enemy soldier or kid, for all I know, let loose with some kind of rocket. It hit up high, but was enough to rock the tank enough to cause the road wheel arm spindle (which was still under tension from the torsion bar), to roll over on my leg, which was right on top of one of the center guides.

So I had a hole in my leg, pinned to the track and some idiots shooting at me. The crater from the mine was deep enough to protect me, until we got the spindle up and out of the way. We did that by threading the tow cables over the center track support wheel from the rear. Another tank hooked on to the cable and slowly pulled back lifting the arm. I was glad to get out of there.

My second tour was with 3rd tank battalion in the northern I Corp. from May 3rd through January 11, 1969, at which time I got an early out for college, cutting my three year enlistment by a month.

Background
Written by James Johnson

I arrived for my second tour as a Corporal and after a month in a provisional rifle company working near Quang Tri, I was assigned to 3rd tank battalion.

I was assigned as a driver for Lieutenant Murphy. He was in charge of the 3rd platoon. His father was a General in the USMC and he was following in his footsteps. At that time nobody in his platoon was aware that I was on my second tour, so I was the "Norman New Guy".

We headed out from Quang Tri, (a major ammo supply area, equipped with 175 mm artillery called the Long Tom's). We were headed out to a place called the "Rock Pile" and our mission, believe it or not, was to carry a bill collector to confront a Marine who had fallen behind on his insurance premiums. Nobody could believe that this was happening, so we all figured his company was just trying to get rid of him.

He was afforded the ride to the Rock Pile (I don't know how) so we obligated him by loading him up on the tank and letting him ride on the back of the tank by the storage rack in back of the turret. Now this was a pretty secure area, but he didn't know that. Lt. Murphy decided to scare hell out of him, so as soon as we left the Quang Tri base area he gave the order to "button up" (lock all the hatches) and "lock and load". This freaked the fellow out for he was hiding and ducking on the back of the tank all by himself.

We dropped him off at the Rock Pile and never heard from him again. I hope he got out of there, but I also hope he didn't collect any money.

We then proceeded to Con Tien. Lt. Murphy and I got into several combat situations supporting Kilo and Lima company infantry. At one point we hit a hornet's nest of NVA. They were so close and so many, I was looking face to face with them at a range of six inches. They were looking through the periscope windows of the cupola. They were trying to throw grenades down the gun tube until we had to call napalm air strikes on our own position.

This did the trick, and it taught me a lot about having back up for these types of situations. We made it out of that mess and LT. Murphy and I departed company. He, of course, took on a new platoon, and I was promoted to Sergeant. I was made a heavy section leader of the 3rd platoon. I never did hear of him again. I am sure he made it up the ranks quickly.

We were sent to an outpost north of Quang Tri, called "Camp Little John". It was just an area on the map with nothing there, just jungle. We were on call to support infantry units for a month or so. At this time we had a platoon sergeant whose name I do not remember. There were five tanks in our unit.

I had some really good people in that platoon. Corporal Cook was my driver and he was good - there is a knack to driving one of those beasts. LCpl. Pierce was my gunner and he was also excellent. He was from the Okefenokee

Swamps and joined the Corps at 17 years of age. He still amazed me how he could find goobers (wild peanuts) in the jungle.

Another excellent Marine and friend was Otis "Marty" Martin. He was on another tank as a gunner. He was from Texas in the town where they made "Yellow Jacket" boats. I remember this because I had one. Marty decided to become a Catholic while he was over there and asked me to become his godfather. I was proud to do so.

First Story
Written by James Johnson

Date approximately October 1968.
Operation Thor (I believe, not sure)
Four days of hell with a little comedy worked in.

A large scale battalion type operation was scheduled in late October of 1968. It involved a massive sweep of the area from Quang Tri up to the DMZ. The main forge was to sweep north from the coast line and cover an area of about 25km (25 clicks). The NVA had been building up in this area for some time. How close they were to us was soon to be discovered.

Our tank unit at Little John was to support three companies of infantry who were with the main force. We were to sweep west from Little John and then head north in conjunction with the main force. The day came and we all mounted up and headed west.

We had good communications with each company of infantry, including each platoon leader. The tanks had exceptional radio capabilities so we could monitor at least four frequencies at a time and transmit on them by simply switching the main transmitter with a knob which we preset to quickly access these channels. Sometimes it was hard to keep up with all the transmissions coming at once, but it was a lot better than operating without knowing what the other units were doing.

The infantry spread out and moved in front of the tanks. We couldn't have gotten 90 meters (100 yards) out of the camp when we hit resistance that I had never seen before. Almost immediately, two tanks were hit with RPGs. The platoon sergeant was hit in the eyes with shrapnel and blinded. The second tank, commanded by a sergeant from Hawaii, was hit knocking out his ability to transmit on the radio. He could receive. The platoon sergeant was MedEvaced and replaced by one of his crewmen. A grunt was grabbed to act as loader. I talked to him later, and he very explicitly conveyed the message that he first thought tanks were an easy ride compared to grunting it, but he would never get near one of those clanks again. They sure provided protection from small arms, but he found out that they also drew fire from the bigger stuff.

What we had hit was what I considered to be the main NVA force. There were a lot of them and they were armed to the teeth with RPGs and 57 mm recoilless rifles. They were also supplied with artillery support from the north. Just a week prior to the operation they had scored a direct hit on the ammo dump in Quang Tri. The fireworks from the explosions lasted two days and caused a lot of damage. I bet the FO (forward observer) who called in the artillery got the highest metal that the NVA could give!

Looking out to the west we could observe that they had set up in shallow trenches. The trenches were aligned in a north-south configuration, spaced approximately 45 meters (50 yards) apart. This was a time for air strikes, but they were all tied up with the main force headed north.

The first trench was attacked by the infantry and there was a lot of hand-to-hand combat at first. The tanks were right there with them, so close that we couldn't lower the main gun and 30 cal. machine gun into the trench. I could, however, provide fire with the 50 cal. and did so at a range of 20 feet or so.

While doing this, I turned my head toward the south from my position in the tank commander's hatch. I then saw the dreaded RPG launch from the trench and watched it head straight for me. It seemed like an eternity before it hit, but when it did the air cleaner box on the left side was blown off and shrapnel flew. A piece somehow found its way up my flak jacket where it tore a pretty good hole up the middle of my back. It was hot and caused blood to start flowing pretty heavily. I was young and really pumped up now. We pulled the tank back from the line and the bleeding stopped. It was more of a scratch than a wound.

The infantry was in contact by radio, and they were taking casualties in this close fighting. I asked them to pull back, so the other tanks could provide cover for me while I used my tank to physically run over Charlie with the tracks of my tank. We buttoned up and proceeded to cruise down the two hundred yard long trench with the right track in the trench and the main gun and both machine guns blazing. My driver, Corporal Cook, was as much of an expert as anyone I have known, and guided the tank as fast as it would go down the trench. He accomplished what I had in mind. He threw up when everything was over.

During all this, I kept getting calls from the supposed commander of operations. He was in the rear, close to our departure point. It was getting to be nightfall at the time and a meeting was called. The commander was a Major, but the first time I saw him I couldn't tell. Instead of U. S. insignias on his collar and cover, he wore Vietnamese insignias. This guy was a glory hound. He wanted to make a name for himself. He balled me out for getting the tanks in front of the grunts. He then rambled on with all kinds of tactics he had learned in OCS including a "herringbone" type of attack formation. I had never heard of a herringbone formation.

When we walked back with the captains from the infantry unit, we all agreed he was crazy and would get a lot of Marines killed.

The second day started with a bang after a whole night of small arms fire and a few RPGs from the opposing trenches. We were working west toward the second trench when the NVA decided to attack heavily. We positioned the tanks in front of the infantry, despite the order from the Major with the Vietnamese insignias. We used a lot of "canister" rounds (almost like a shotgun shell with a lot larger pellets), and kept a steady flow of HE (high explosive) rounds, flying toward the immediate trench and out further to the other trenches. All the time the idiot commander was on the radio telling me to pull back. We didn't have a choice except to keep the attack going. We were outnumbered and were receiving incoming artillery fire from the north. The grunts were advancing with us, picking off people with RPGs. My response to him and from the infantry companies was, "You're coming in garbled, cannot understand, over". Hey, it worked!

We continued on, and went through three more trenches before the day was over. It took two days to go about three hundred yards. We had to call in ammunition via choppers twice during this period. Watching the NVA move to the north, we pulled into position for the night. Everything was fairly quiet. Backing a tank into position requires a "ground guide", someone in front of the tank to hand signal the driver as to which way to turn.

I watched as my godson, Marty Martin, guided his tank into position when the tank hit a large mine. The blast picked up Marty and threw him a good ten yards. He didn't get hit by shrapnel but the blast twisted his torso so bad that we all believed he had a broken back. He couldn't move his lower body at all.

He asked me what was going to happen. I sure didn't give him any encouragement, as I told him he wouldn't walk again. A chopper MedEvac was called in and as I watched I kicked myself for being so blunt and not offering encouragement. For God's sake, he was my godson. He surprised me a month later when he came walking back for duty, happy that it was only nerves that healed themselves.

The next day we were hit pretty hard from the north - not any entrenched NVA but a lot of artillery from the north. We could only hold our position and try to locate the guns that were lobbing the shells in on us. During this day, my gunner had a case of severe diarrhea. Since he couldn't really go outside the tank because of the incoming rounds, he had no choice but to let loose in his trousers and go inside. I never knew a fellow could go that much. He must have filled a good inch in the deck of the turret. My gunner LCpl. Pierce, put on his gas mask because he was down in the middle of the mess and couldn't get any air. To add to this, the deck was pretty well filled with empty brass from the machine guns. It was a smelly mess! We had to MedEvac my loader, so we were operating with a three man crew.

By this time my company commander, who was monitoring our actions from his position in the main force, decided to send a staff sergeant from the rear in Qung Tri to take over the command of our unit of tanks. He had the information that the platoon sergeant was blinded and one tank was not able to communicate. He also probably knew I was blowing off the looney major in charge over the radio. By this time they had come and got the crazy major. They eventually sent him back to the states. He did however, have time to issue a court marshal offense against me involving disobeying orders. Hell, he was coming in garbled! I did have to answer a few questions later, but nothing ever came of it.

The staff sergeant arrived by chopper during a barrage of incoming artillery. It was pretty big stuff, probably 155 mm, and had left our surrounding area peppered with craters from ten to 20 feet wide and 3 to 5 feet deep. Well, here comes the chopper and literally drops him from the rails. He lands about 30 yards from us and positioned himself in one of the craters. Artillery is coming in while he seems to hop from crater to crater to get to the tank. He was an older Marine from Korea and was spending his tour in the rear as an office clerk. I don't think he wanted to do this at all.

It took him a few minutes to make it to my tank. He climbed up on the tank from the rear and made it to the loaders hatch. The artillery was coming in pretty heavy and as he started down the hatch a shell exploded next to the tank. This blew him in and he landed pretty hard in the brass and layer of diarrhea left over from my loader. This was not a good scene, but I welcomed him aboard and offered him a towel I carried on the radio rack.

During all this, the infantry had located the position of the enemy artillery. It was located several miles to the north and was comprised at three different locations. We were firing back as best we could, acting as artillery pieces to try to quiet the enemy bombardment. I was getting calls from all three infantry companies and was also calling in air strikes on the enemy positions. The earphones were buzzing, and as the staff sergeant wiped off his com helmet and put it on, he looked at me, shook his head, and said "You take over, sergeant". So I did.

That day we advanced to the north and were actually chasing the NVA to the north. By this time we had full air and artillery support, which was diverted from our main force. It seems we hit the main force of NVA, who were closer to Quang Tri than anybody knew. I had never seen such firepower delivered to one area for so long a time. It was amazing and sure pumped up the adrenaline level in all of us. By the end of the day we had succeeded in

142

driving the NVA back home and away from Quang Tri. That night we had a little rest.

The next day everybody loaded up for the return trip to Little John. Things were quiet. The tanks carried as many of the wounded Marines as possible on the back. They were Marines that didn't necessarily need MedEvac, but were injured enough to not be able to walk back.

On the way back, the infantry companies were marching in columns. As we passed by these columns with the wounded aboard, the grunts raised their hands in salute to the tanks and gave a cheer. It really made me feel good. The tanks were pretty well shot up, missing fenders and spotlights, and my tank missing an air cleaner. Although I had wrapped a blanket around what was left of the air cleaner, the engine started to give out and was throwing billows of black smoke out the back. It made it to our destination, where we loaded it onto a Mike boat, which brought it back to the rear.

After we loaded up our tank, I was confronted by a Colonel, my company captain, and my old tank radio instructor, Gunny Woodward. I didn't know the Gunny was over there and was surprised to see him. They were there because of the ruckus with the looney major with the wrong insignias. There was a tent set up and we all went in. They explained what the whole situation was about. I was a little scared because I did blow the major off and could really hear his frantic radio transmissions.

After a formal reading of what this was about and what the charges were, the procedures abruptly stopped as a bottle of whiskey appeared from Gunny Woodward's field jacket. We all had a shot and went about our business. As I said, the looney major had already been ordered back to the states.

The last I saw of my tank was at the tank headquarters in Qunag Tri. They had stripped it of anything useable and buried it into the dirt. I figure nobody wanted to fix her back up. Hell of a way for Alpha 33 to go down.

Tanks were really used in Vietnam. They saved a lot of lives. One of the things that I'm proud of is that no one in my platoon was ever killed, some were injured, but none killed.

Sergeant James F. Johnson 2265467

Tanks of Afghanistan and Iraq Wars
Written by Clyde Hoch

The M1A1 Abrams was produced in 1980's and is used today, in different militaries around the world. It was a radical design difference from the older tanks. It carried a lower, smaller silhouette, with composite armor. Armor was depleted uranium, ceramic, composite plastic and Kevlar. It is considered a heavy tank.

The M1A1 had a radical difference in optics, fire control system and crew safety.

When the M1 originally came out around 1978 it carried a 105 millimeter, rifle barreled main gun. The M1A1 carried a 120 millimeter smooth bore main gun of German design. It carried 42 rounds. It was also equipped with a 12.7 millimeter gun and two 7.62 machine guns.

The M1A1 was 32 feet long, 12 feet wide and 8 feet high. It carried a crew of four - tank commander, gunner, driver and loader. It weighed in at 65 tons combat loaded.

The engine was a multi-fuel turbine and runs on kerosene, diesel or gasoline. It produced 1500 horse power. It had a fuel capacity of 500 gallons and a range of 265 miles. Top speed is around 60 miles per hour. It can fire the main gun accurately on the move.

M1Ai Abrams firing its main gun

M60A1 also a Patton type tank, was meant to replace the M48A3. It had a similar hull. The M-60A1 was the last tank manufactured in the U. S. to have a homogeneous steel hull. It was also the last to have an escape hatch in the driver's compartment in the belly of the tank.

The M-60A1 was used by the U. S. military from 1961 to 1997 in various variations. The turret was made longer in some. The main gun was changed to a 105 millimeter in later forms of the M-60. 15,000 were produced in total.

The M60A1 weighed in at about 51 tons. It was 22 feet 9 ½ inches long. It was 11 feet 11 inches wide and 10 feet 6 ½ inches high.

It was powered by a Continental V-12 air cooled twin turbo diesel engine. It could carry 385 gallons of fuel and had a range of 300 miles, at a top speed of 30 miles per hour.

M-60A1 pictured. Some notable differences between an M-48A3 and the M-60A1 were a straight front of the slope plate where the M-48A3 had a bow slope plate. The M-48A3 had a blast deflector on the end of the gun tube. The M-60 had less return idler wheels than the M-48A3.

Donald C. May Senior "C" Company Vietnam 1966

Written by Clyde Hoch

Photo by First Sergeant Rick Lewis, who said
Don Sr. was an outstanding and hard working Marine.

Donald C May Junior

As I said earlier, four of us came into 2nd Tank Battalion at the same time - Gary Young, Robert Alexander, Donald C May and myself. We were all new, the same age and experience level. We spent a total of a year and a half together. We went on a Mediterranean cruise for six months on a small ship. We were always together and became very good friends. You could fill a book with some of our exploits on the Med cruise.

After a year and a half together Gary Young was sent to Vietnam. Alexander and I went on another Med cruise. I totally lost track of Don C. May Senior. I searched the net for any information I could get on him, with no success. John Wear printed a short section in the USMC Victnam Tankers Association newsletter in search for him.

I finally got a phone call from a Michael (Muddy Water) Water. He said Don SR. was a Corporal in 1st Tank Battalion, "C" Company, 2nd platoon in 1967, South of Danang. The same area I was in a year later. He said Don was TC on C-22, the tank Michael was a driver on.

I heard Don passed away from a heart attack, on a fishing boat after he got out of the military. He died what he liked doing.

Don was a good guy. I remember him telling me he was from Buffalo, New York. One night after drinking way too much, Don started to slap box with me. I was not in the mood and struck him hard. He kept at it and I kept hitting him hard. The next morning I was surprised that he had no marks on his face. It didn't faze him and he never

149

commented on it. I was happy he was alright but disappointed in myself for not being able to punch hard, in my drunken state.

My daughters were searching through my attic and found three dog tags. I looked at them and put them down. Later I thought, "Why, would I have three?" After closer inspection, I realized one of them was actually Donald C May Senior's dog tag. It reads May, D. C. (serial number, blood type, USMC (gas mask size) S (religion) Catholic. How it came into my possession, I have no idea.

After his military commitment, Don married and had a son, Don C May Junior, in Richmond, Virginia. As Don Junior was growing up, he was always interested in the military. Soon after graduating from high school Don Jr enlisted in the Marine Corps, like his father.

Don Jr became an MP (military police) and guarded Iraqi prisoners during 1991 Gulf War.

Don Jr was promoted to Staff Sergeant and became a tank commander, like his father. Don Jr served in 1st Tank Battalion, 1st Marine Division.

On March 25, 2003 during operation Iraqi Freedom, the tank he was commanding was in a huge column of tanks, on the rush up to Bagdad. They were traveling very fast and the dust was horrible, from all the vehicles. They had been awake for over 36 hours and it was around midnight. They were approaching a bridge on the Euphrates River, near Nasariyah. The main column of tanks veered to the right. Somehow Staff Sergeant May's tank missed seeing the other tanks turn.

The tank plunged down an embankment and landed upside down in the deep river.

Because of the lack of sleep, horrendous dust and raging battle noone noticed a missing tank until the next day. They back tracked until they found where the tank had gone over an embankment. After pulling the tank from the river, to their dismay they found the whole crew died on impact, there were no survivors.

Don Jr had a loving wife Deborah, a daughter, Mariah, and two sons, John and William.

I am sure Don Sr was as proud of his son as Don Jr was proud of his father. Both were tank commanders in two totally different tanks and very different wars. Both served with honor and dignity.

According to First Sergeant Rick Lewis, Don Sr. was an outstanding hard working Marine. He served with Don Sr. near Danang in 1966.

I was proud to know Don Sr. I am very privileged to do this story on both fine American fighting men.

Today there are three tank battalions in the USMC. 1st, 2nd and 4th which is a reserve unit. Each battalion has four line companies and a headquarters company. Each company has four platoons. Red is 1st platoon, White is 2nd platoon and Blue is 3rd platoon. Black 6 is the Commanding Officers tank and Black 5 is the Executive Officers tank. Each platoon is made up of four tanks, split into two sections. Blue 1 is the 3rd platoon Commander. Blue 1 Golf is the gunner on 3-1 or Blue 1.

In the streets of Fallujah, the streets running east to west were given female names. The streets running north to south were given male names.

Captain Jeffery T. Lee 1802 / USMC
Written by Capt. Lee

All photos in this chapter were supplied by Capt. Lee

Jeffery Lee enlisted in the Marine Corps from Spartanburg, S.C. in August of 1988 and graduated Marine Corps Recruit Depot, Parris Island S.C. November 1988. After initial MOS training at Fort Knox, KY as an M60A1 Tank crewman, he was stationed with 2nd Tank Battalion, Camp Lejeune, N.C. from 1989-1992. While there he conducted a unit deployment to Okinawa, Japan and a deployment in support of combat operations in Desert Shield and Desert Storm. Prior to transferring from Camp Lejeune to Columbia, S.C. he attained the rank of Corporal (Cpl). In Columbia, S.C., Cpl Lee was assigned to the Inspector-Instructor Staff (I&I), for D company, 8th Tank Battalion, Marine Reserve Forces from 1992-1995. While stationed with the I&I staff, he transitioned the Reserve company from the M60A1 Tank to the M1A1 main battle tank. In October 1994, Sgt Lee was reassigned to 2nd Tank Battalion for duty with Bravo Company. His duties during this assignment included Tank Commander, Section Leader and Platoon Sergeant before transferring in June 1997.

In 1997 he was promoted to Staff Sergeant and reported to Marine Corps detachment in Fort Knox Kentucky to train tank crewman. In1999 he was a student at the staff NCO program. May 2000, he was promoted to gunnery sergeant. In April he went to officers candidate school. In December 2002 he was commissioned to second lieutenant. In August 2012, Capt Lee was assigned as Company Commander, Company A, 2d Tank Bn where he currently serves. Capt Lee's personal decorations include the Silver Star, Purple Heart, (4) Navy Commendation Medals, (3) Navy and Marine Corps Achievement Medals, (2) Combat Action Ribbons, (4) Good Conduct Medals.

Capt Lee is married to the former Jolynn E. Yates of St. Louis, MO. They currently reside in Camp Lejeune, NC with two of their 4 children: Carolina (16), and Jace (4). The other two Hope (21) is married to a Marine stationed at Cherry Point and Donovan (19) is a Marine attending Public affair school at Fort Meade, Md.

Capt Lee in Iraq

DAAB 1989

The Marine Corps M60A1 Rise Passive Tank, Breaching Operation, Operation Desert Storm

Written by Capt. Lee

In August 1990, Delta 1st Armored Assault Battalion (DAAB Company), was deployed on a Unit Deployment Program (UDP) to Camp Fuji, Japan. The Company received orders to deploy in support of Operation Desert Shield and subsequently was ordered to Okinawa, Japan to receive all necessary desert combat gear, as well as medical and mobilization screenings. Around the beginning of September, the company deployed to Saudi Arabia, where we would be attached to 3rd Tank Battalion, 1st Marine Division as Delta Company. Once we arrived, we immediately began to receive our M60A1 Rise Passive Main Battle Tanks. The Company then began our acclimatization process by staying at an American Oil workers camp for 30 days. While there we rotated forces every 3-5 days from the field to the camp. At the beginning of November 1990, our company moved all forces to our tanks and combat trains in the field, and that is where we continued to live until March 1991.

From the time I arrived in Saudi Arabia, I had spent about nine days total at the oil workers camp that was assigned to the Marine Corps, and the rest of the time I would live on my M60A1 Rise Passive Tank. From November 1990 until February 1991, 3rd Tank Battalion would conduct movements every 2-4 days so not to be stationary for too long. We would also conduct rehearsals in breaching operations at the company and battalion level so that we would be very proficient at breaching the mine fields that were to come. I can tell you without exaggeration, my company alone conducted over 20-30 mine field breach rehearsals throughout December and January. As a Lance Corporal, and the driver of the platoon Sergeants tank, I knew exactly where I needed to be in a breach, where I needed to go and what I needed to do once the breaching operations started. I was very confident in what I believed our capabilities were and what we would be doing. I was not confident that we were properly informed.

Once the Air Campaign started in January, we knew it would not be long until the ground war would start. As we moved around, rehearsed, and prepared for the ground battle, we mentally hardened ourselves to what was ahead. We would sit around and talk about what was coming, then talk about our homes and what we were missing. I can still remember lying at nights on the back deck of the tank listening to my CD player and staring at the stars, they were so clear and bright. I would sit and wonder what the fight would entail, what my family back home was doing, what I would be doing if I were still at home or what my life would be like if I had not even joined the Marine Corps. Then I would fall asleep and awake to another day of rehearsals, movement or conducting preventative maintenance of the tank.

In December, the 3rd Tank Battalion Commanding Officer was notified the Marine Corps had purchased M1A1 Main Battle Tanks. Headquarters, United States Marine Corps knew that 3rd Tank Battalion had deployed with the

M60A1 MBTs after taking them from the Maritime Prepositions Ships (MPS) and wanted the battalion to transition to the M1A1. The Battalion Commander asked the Marines of 3rdTank battalion if they wanted to fight with the M60A1 or transition to the M1A1 in midstream. Without hesitation, the entire battalion unanimously agreed to keep the M60A1 instead of spending two weeks training on a tank we had never seen, used, and were expected to fight with at any given time. For the Marines at that time, it was an easy decision. However, knowing what I know now about the M1A1 Abrams tank, it would have been a fast and easy transition with even more devastating effects on the enemy.

Toward the end of February, it was apparent the coming ground war was near. We were moving more northward from Saudi Arabia and aligning the battalion alongside of foreign forces. Also, friendly U.S. Army forces were maneuvering in and around our location. We were getting our engineers, Forward Air Controllers (FACS), and Forward Observers (FO). As a Lance Corporal, I could see things happening all around me, lots of movements and personnel changes. It just made sense we were about to fight. On the evening of 23 February, we were given the operations order and told what we were expected to face in the coming days. The battalion would face off against a division of tanks and personnel carriers. We were then told we would be moving close to the Line of Departure (LD) and we would prepare to attack. At that point, we were told to conduct Preventative Maintenance Checks and Services (PMCS), check our gear and equipment and then prepare for movement. We took down the cammie netting, rolled it up and put it away on the tank. Finally, three months of using cammie netting every day and we are now possibly done with it.

At about 1400, we began our movement to our attack position. When we arrived around 1700, Armored Combat Earthmover (ACE) were busy digging hasty tank fighting holes to protect the tanks from Indirect fires (IDF) from Iraqi artillery and long range mortars. We were instructed to eat dinner and put on our MOPP (Mission Orientated Protective Posture) gear for possible chemical attack and rest before LD. At around 0400 on 24 February, we began our movement to LD. Our mission was to be at the first mine field at daybreak. The commander believed the sun would rise behind the Iraqi tanks and that would silhouette them on the horizon giving us a better advantage in the fight. We arrived while it was still dark, around 0515. We set up on the mine field as we had done in our past rehearsals waiting on daybreak. This was due to only having passive sight capabilities on the M60A1 Rise passive tank, the same capabilities as the Iraqi tanks we were fighting. A passive sight is like that of a night vision goggle. As the sun rose, what looked like a regiment of Iraqi tanks were backing out of their defensive positions. Also, Iraqi soldiers were running around on the opposite side of the mine field. There were also high towers that lined the mine field as what seemed to be an early warning to our approach; it seemed to not have worked. The Iraqi tanks looked as if they were conducting a withdrawal, but a day too late.

Task for Ripper, the task force 3rd Tank Battalion was assigned, had set online and was prepared to conduct their attack. The order came across the radio to conduct the breach and secure objective 1. The first tank shots rang out from the M60A1 of the battalion and the assault began. Iraqi tanks were exploding on the horizon as the turrets of the T54/55 tanks were blown from their hulls. AAV and Tank machine guns were rocking at the Iraqi soldiers running around the mine field and climbing the towers on the mine field. Marine Corps Ah-1 Cobra helicopters were flying in and out of our formation, roughly 100 feet off the deck, firing hellfire rocket's at tanks on the horizon and at the towers. Since my tank was a proofing tank for a breaching lane, we began our move behind the mine plow through the lane. Once we were through the lane, we immediately broke left and started our movement toward the enemy on the horizon. At this point, the enemy tanks seemed to be about 1500-1800 meters from our position. I could see the enemy tanks firing and the tanks rounds hitting about 50-100 meters short of our tanks as we continued to move. I could see and hear the M60A1 tanks firing and enemy tanks continuing to explode to my front. Within an hour Objective 1 was secured and all Iraqi tanks were neutralized or destroyed. We had no damage to any of our companies M60A1 tanks while enemy vehicles were still on fire all around us.

While we were approaching the destroyed tanks and personnel carriers (BMP 1), I could see burned and charred bodies melted to the tops of the turrets and lying all around the area. The turrets had been blown off the hulls of the tanks and the fires were so hot that enemy soldiers had not had time to exit the vehicles. The battalion did not stay at this position long. We only stayed long enough to have conducted consolidation, reorganization, and any resupply/refuel that was needed. We then received orders to move to the next mine field.

The second mine field was breached just as quickly as the first mine field. The enemy was waiting on us this time. However, the T54/55 tanks of the Iraqis could not range us at 2000 meters as they had thought. We again started to destroy enemy tanks as we breached the second mine field. The second mine field looked the same as the first with towers, soldiers manning the towers, and a defensive position over watching the mine field; However there was a difference. Once we secured the objective, the enemy began to surrender. As we sat in a defensive posture, we pointed for the Iraqi soldiers to start walking south to link up with the infantry moving up behind as 2nd echelon forces. Also, we began to receive the first indirect fires (IDF) of the war. IDF rounds were impacting pretty close to the tanks. We had to move our positions left and right to keep the IDF from becoming accurate. One of the tanks in the platoon was towing a line charge trailer for the engineers. That tank took a mortar hit to the right front fender. No damage was done to the track for mobility purposes, but the front fender was blown off. The company called Marine Corps artillery on the suspected IDF location and all enemy fires ceased. The plan to breach the Iraqi mine fields had gone faster and much better than any planner, commander or even the enemy force had imagined. The breach was so quick that the enemy did not have a lot of time to react or to conduct a counterattack. If the enemy had a plan, they did not execute that plan at all. 3rd Tank Battalion, pushed the enemy back on their heels utilizing combined arms, speed and shock effect and never let them catch their breath during these breaches.

Throughout Operation Desert Storm, the M60A1 Rise Passive Tank performed superbly. It out ranged the Iraqi T54/55 tanks and BMP 1 of the Iraqi army. They had very minor maintenance issues during the entire four days of the war; However, the tank did have some combat limitations, i.e. passive sights for the driver during blacked out instances while oil fields were being burned, no ability to shoot at night without some starlight and very slow speeds across desert terrain. Ultimately, the M60A1 MBT outmatched the Iraqi tanks forces.

Captain Lee upper left on 3-1

Company A, 3rd Platoon
Operation Phantom Fury (Al Fajr)

Written by Capt. Lee

This road started for me on October 2, 2002 as a Gunnery Sergeant. On this day I reported to Officer Candidate School (OCS) in Quantico, Virginia after being selected for commissioning by the Enlisted Commissioning Program (ECP). At that time, I had 14 years in the Marine Corps and had served on the M60A1 Tank in Operation Desert Shield and Desert Storm. I then transitioned to the M1A1 Main Battle Tank (MBT) in 1992 while I was stationed in Columbia, South Carolina training a reserve company (Delta Company, 4th Tank Battalion; which has subsequently been disbanded). After OCS I attended The Basic School in Quantico, Virginia from January through July 2003. TBS is the school all Marine Corps officers attend to learn what it is to be a Marine Corps officer and a Second Lieutenant of Marines. It is also where every officer earns their Military Occupational Specialty (MOS) for their career in the Marine Corps. My Staff Platoon Commander (SPC) had to fight for me to have the armor MOS as we only had two slots for armor officers and the top slot had been taken by a Marine that was one ranking higher than me. Normally, there is one tank slot in the top of the class and one toward the middle of the class. We both placed at the top, so there had to be a special request sent to the Commanding Officer of The Basic School and a very good rationalization on why this matter was special. I am not sure what was said, I just thank God everyday that I was approved for the MOS and that I have had the

163

opportunity to serve in the armor force. My assignment in Knox, Kentucky, was Company A, 2d Tank Battalion in Camp Lejeune, North Carolina in June 2004 and that assignment finally led me to Operation Phantom Fury/Al Fajr during November 2004.

The Turnover
Written by Capt. Lee

At the end of September 2004, Company A deployed in
support of Operations Iraqi Freedom 2-2 to the Sunni
Triangle just outside the city of Al Fallujah. At the time, I
had no idea what was going on since I had either been in
school or training my platoon in preparation for the
deployment. However, during my first turnover patrol with
the outbound platoon commander, it became very apparent
this deployment would be a tough fight. During the initial
patrol on the outskirts of the city, our tanks began to take
RPG fire. None of the RPG fires were accurate nor did
they hinder our movement. However the fire did facilitate
a mental hardening in what we would be dealing with in
the near future and was preparation for what was to come.

Tanks in Shaping Operations
Written by Capt. Lee

Throughout October, our company was assigned to conduct probing patrols both in the day and at night. Our patrols would take us initially to the western side of the city, where my 3-2 tank took an anti-tank mine to the right side track. We also came under accurate RPG fire that resulted in me calling the M88 recovery vehicle to pull the 3-2 tank back to Forward Operating Base (FOB) Fallujah leaving the track in place. Another patrol took us to the southern side of the city where we received no contact. During our patrol to the northern side of the city, we were also tasked with destroying Texas barriers (large concert barriers) that were placed to block the roads into the city. As we drove the northern sector of the city, every street had Texas barriers placed at their beginning. As we stayed approximately 1200 meters out, we engaged and destroyed every barrier. All of these patrols and engagements were meant to flush out the enemy while UAVs were flying overhead and to set conditions for an attack. During this time as well, coalition forces were sending messages for civilians in the city to leave before Marine Forces took the city by force.

Phantom Fury
Written by Capt. Lee

After dinner on 2 November, my company commander assigned my platoon to support 1st Battalion, 3rd Marines for initial entry into the city of Fallujah. I was informed that planning the mission would start that night and that I needed to be there to assist in the planning of the use of tanks for the upcoming fight. As I was still a 2nd Lt and had never been part of planning a major operation, I was a little unsure of what I would be doing. As I showed up to the Operational Planning Team (OPT), I was immediately engaged in the capabilities and limitations of the M1A1 Abrams tank and any special considerations that may be needed. I was able to give my honest input to the staff and to facilitate how tanks would be used by 1/3 during Operation Phantom Fury. I felt as though I was contributing to the success and to the safety of infantry Marines in the fight. The planning became a lot easier when the infantry battalion no longer had to worry about the logistics to support a tank platoon. My company commander decided to use all his resources to establish a Logistic Supply Area (LSA) for the entire tank company so the infantry did not have to worry about supplying tanks with ammunition, fuel or chow. We also had a place to conduct maintenance when needed. This seemly small event was astronomical in the large scheme of the fight. This concept allowed the infantry to focus their attention on their resupplies, CASEVACs, and refueling operations with what they had on hand instead of requesting even more outside support from other agencies.

After 24 hours of planning it was confirmed, tanks would lead the entire fight into the city. Now we would spend the next two days training tank and infantry integration to facilitate the fight in the city and how to provide tanks with security inside a built up area. After the planning, I brought my tank commanders and crews together to rehearse the tank/infantry integration training in an effort to ensure every tank crewman was teaching the same information in the same manner. Each tank section would teach a company of infantry Marines and rehearse with them for a four hour time block. Then we would rotate to another company until we were complete with the training. After the training, the platoon would be attached to a company for initial entry into the city of Fallujah and then would shift companies based on main effort shifts and phase lines. Each tank commander was informed that he would attend each infantry company's operations order to answer questions about tank capabilities and limitations and also to know what was expected of each tank by the company commanders.

The following was an exact time line leading up to A-hour (attack hour).

Phantom Fury Timeline:

D-3 (4 Nov)

Company Orders development

D-2 (5 Nov)

Company orders/ Battalion ROCWALK/ 1800 Regimental ROCWALK

D-1 (6 Nov)

0900 Battalion Rehearsals / Tank Maintenance day

D Day (7 Nov)

0800-1200 Pre-combat checks/Pre-combat inspections

1200-1500 Staging

1500-1600 COMM checks

1600 Chow/rest plan

D+1 (8 Nov)

0300 Quartering party gather

0400 Quartering party departs

0500 Comm checks / serial one departs

0700 Serial two departs

1000 Stage at Attack position

1900 A-Hour

On 7 November, my tank platoon was prepped and set for movement to stage at the 1/3 staging area. However, I knew we would not get an opportunity on the Marine Corps birthday to read the Commandants message or have a piece of cake. So, once we had the tank set for movement I had my platoon sergeant call all the Marines over to my tank. I then gathered them around and discussed the coming fight. I explained that if we were going to miss a Marine Corps Birthday, there was no better way to celebrate then to be fighting for freedom and all the Marines to our left and right. I then read the Commandants message, had some MRE pound cake, sang the Marine Hymn, said a prayer for the platoon and then mounted the tanks, received the ready condition from each tank and then moved to the staging area.

The above was the timeline followed by the regiment. Tanks were part of serial two and set in the attack position by 1200 on the 8th. With a regiment all moving at the same time and using the same route, a back-log of traffic builds up on the routes to the assault position. We were essentially in a traffic jam until we reached our deployment lane. At which time, we deployed on line to the assault position to wait on the assault.

Engineers were the first to move toward the city. They were going to cut breach lanes across the rail road tracks and into the city. A suspected minefield was set just outside the city to prevent an assault or at least to try and slow it down. So when 1900 came, the engineers moved to the breach lane and commenced to conduct the breach. At around 2100, I ordered my tanks to begin to move closer to the breach site. However, when we had moved just under a set of high voltage power lines, we began to receive 120mm mortar fire from the city. This fire was very accurate, forcing us to move even closer to the city to get out of the fire. Once we stopped, my 3-3 tanks engine blew. The downing of that tank took an entire section of tank out of the fight, as my 3-4 tank would have to pull the 3-3 tank to the LSA for repair or replacement. My 2-2 (wingman) tank and I continued to wait for the breach lane to open, but by midnight we had to find another way in. The identified breach lane had a large berm that needed to be built up so all tanks and wheeled vehicles could cross the railroad tracks. Since the berm at the railroad tracks was so tall, it was taking a lot longer than expected to create the breach lane needed for tanks. Instead of waiting,

I took both my section of tanks to our west and linked up with a U.S. Army unit, Task Force 2-2 and used their breach lane to enter the city. I then moved back east on a hard surface road until I reached my entry point to the city. I waited at the entry point until I could link up with my infantry squad which was moving across the railroad tracks at the breach site as we were getting set. Once the link up occurred at about 0100, we launched our assault.

As we entered the city we began to take small arms fire from the tops of the building. The infantry was trying to find places to get out of the fire, but there were no hiding spots at that time. So instead of keeping the infantry in the open, I moved up quickly to an open area with concrete fencing so the infantry could find a better vantage point or to get higher in the buildings and to destroy the enemy. As my tank section came into the open, there was a mosque to my right flank (east). From the mosque, we began to take RPG and medium machine gun fires. I ordered the infantry to take cover and I then moved my section toward the mosque to engage the threat. Once the enemy fire was extinguished, I ordered the infantry up to clear the mosque. My tank section then set a hasty defense oriented south. By this time, it was around 0700. The sun was up and the infantry was getting ready to begin another push to battalion objective two. Before we started our move around 0800, the infantry came under intense fire from the West. My tank section could not fire in a western direction due to friendly forces operating in close proximity to us. So, we coordinated with our adjacent units and neutralized the enemy to the west. Around 0900, we began our movement south toward objective two. As the tanks moved at the speed of the infantry, we would receive sporadic sniper fire from different locations.

At one point, my gunner identified a sniper position and destroyed it with 7.62 coaxial machine gun fire. The rest of the movement was mostly the same; A little small arms fire and sporadic sniper fires. Once we reached Dave field (objective 2 and an old soccer field that had become a cemetery) the infantry began to set in a defensive posture oriented south. I then set my tank section in a hasty defensive position oriented south overlooking Dave field.

3rd platoon in attack position

Resupply

Blue 2 enroute

Let loose with the big one.

Grunts and tanks working together

This position gave me observation for at least 800 meters to my front across the field and 200 meters to my flanks.

At this point, my tank section had been operating for about 39 hours on one tank of fuel and the ammunition with which we started. After setting in, I called back to my bravo section (3-3 and 3-4) and learned the section was moving back to my position but hit an anti-tank mine on the way in. They had to go back to the LSA to fix the track in a safe location. Now it was about 1800 on the 9th and the bravo section was within eyesight of my location when the 3-4 tank (platoon sergeants tank) took an anti-tank RPG though the engine compartment fuel cell. It did not catch fire, but fuel was pouring out the bottom of the tank. The bravo section then turned around and went back to the LSA. That evening, my section sat overwatch on Dave field and during infantry patrols. Throughout the evening and night, my section engaged dozens of insurgents moving in and around Dave field and Phase Line Fran (a main hard surface road running east to west though the middle of the city). On the morning of 10 November, my bravo section reached my location and completed a swap with my tanks. At this time, my tank section had been running for over 50 hours on one tank of fuel and one load of ammunition. I was not sure I could make it to the LSA for fuel.

During the last 50 hours, I knew that I could not leave the infantry without tank support. Every time I would move away or an infantry patrol would move away from the tanks, they would come under intense enemy small arms fire. I did not want to leave the infantry without M1A1 tank support and protection. I took a large risk to stay in support of the infantry, but it was a necessary risk. Now, it was time to get some needed resupplies, refuel, maintenance and rest. We arrived back at the LSA around 1300 to start refuel, upload ammunition, chow and maintenance. After we completed maintenance, I allowed the Marines to sleep for 6 hours. I woke them around 0200 to prep tanks to move back into the fight. I left the LSA around 0300 and arrived at the infantry location around 0400. At that time, we sat overwatch while my bravo section returned to the LSA for fuel and ammunition.

At 0800, I called higher headquarters in an effort to be detached from 1st battalion, 3rd Marines and reassigned to another infantry battalion. At 0900, I was informed that I was detached and reassigned to 1st battalion, 8th Marines. I was told to report to them ASAP at the Mayor's complex as they were in need of support. I moved to that location and linked up with Company A, Capt Aaron Cunningham. He informed me that we would be conducting an attack across phase line Fran and into the teeth of the enemy. He yelled that we would be moving in 10 minutes and I had better be ready or I would get left behind. I ran back to my tanks and gave instructions. My bravo section was not back yet, but we had to move anyway. My company commander and executive officer would also be with 1/8 during this push.

As we moved across Phase Line Fran, I was contacted by an infantry platoon commander who stated he needed fires on the "beach building". He identified his location as the "candy store", a very bright building that looked like a candy store. I turned and found a three story building that had a beach mural painted on the side approximately 600 meters to my front. The platoon commander walked me through where the enemy was in the building and I commenced main gun fire on the "beach building" with two tanks. We put 4 rounds from each tank in the building and all enemy fires ceased from that location. We then pushed down to our assault lane and began the assault. However, through confusion, my wingman (2-2) tank did not get in position behind me on the road and two AAVs pulled in behind me instead. Before I knew what had happened we were moving in the attack.

Seventy meters down the road, there was a little "S" turn in the road. As I approached the "S" turn I was unable to see in any direction due to large homes and high concrete fences all around the turn. Setting directly to my front was a large house. As I turned to look left, another house with a garage was sitting on the turn. These houses prevented me from being able to see my direction of attack. All I could do is push through the S turn with infantry support. As I started to take the left turn, my tank came under intense small arms and medium machine gun fire. The infantry had to find a place to get out of the line of fire (go to ground) as they were in the open against the concrete walls. I pushed a little faster to the right turn to reposition my tank in a southern direction facing down the long axis of the road. As soon as I straightened up, the enemy fire intensified with RPGs, recoilless rifles, and other munitions. My gunner and loader started calling out targets. I halted the tank and gave the fire command, "Gunner MPAT troops in buildings". The Gunner yelled, "Identified". I commanded, "Fire and adjust, caliber 50, on the way". The tank started firing 120mm main gun rounds at enemy defensive positions in the home on the left and right sides of the streets and then the coaxial mounted M240 7.62mm machine gun at troops running on the roads. I was engaging with my M48, caliber .50 machine gun at enemy positions on the rooftops who were trying to get an RPG top down shot on my tank. Since it was only my tank at the location, we were going through ammunition very fast. The gunner was shooting so much coax 7.62mm ammunition, the gun started to jam. The barrel was very red and in some spots it had turned white with heat. The main gun stub bases (stub base is what is left of the main gun ammunition after it fires) were stacking up on the floor.

My .50 caliber was out of ammunition inside the tank and I needed more. I had stowed some on the top of the tank right beside my hatch. So I opened the hatch from an open protected position to the fully open position and then rose to get more ammunition. As I reached a standing position, I was shot in the right bicep. Since I already had the ammunition in my hand, I brought the boxes back in the tanks while at the same time I yelled at the crew that I had been shot. My loader immediately started to find what he could use to patch me up. My gunner took out his KBAR knife and cut off the sleeve of my CVC suit. As the loader patched me up with duct tape, the gunner changed barrels on the M240 coax machine gun and got ready to start shooting again.

While all this was going on, my 3-2 tank came across the radio asking if I was alright. I stated, "My M240 is jammed, my loader is wrapping me with duct tape, and my main gun is elevated. You have less than 30 seconds to get here and assist". It did not take 15 seconds before my 3-2 tank had moved to an adjacent road and came down far enough to bust through the walls of two houses before entering the road to my front as followed by his .50 caliber machinegun.

NOW THAT IS WHAT I CALL A WINGMAN! With my 3-2 wingman, Sgt Ducasse, to my front, we began to move forward so the infantry could access the entrances to each compound, large metal doors, in order to enter the homes to establish rooftop security. Once the infantry from 1/8 gained a foothold on the first few houses, we continued to push until we reached our objective, a school, just after dark. Once we reached the school and had an AC 130 gunship (call sign "Basher") on station, I took my section back to the LSA for resupply, refuel and ammunition. I also was taken to medical to get an antibiotic shot and to get my wound washed out. I would not stay or allow MedEvac as 1/8 was depending on me and my platoon for support to push the attack the next morning.

At this point, I had linked back up with my bravo section. After maintenance was complete, I allowed the crews to get 4 hours of sleep before we headed back. My goal was to be back in the fight by 0500. As we arrived back to 1/8 position, the company called for me. The front of the building they were in was getting rocked by multiple RPG rockets. I was asked to move to the front, find the RPG gunners and destroy them. I had my tanks move to the road and begin to search. At that time, we came under heavy RPG fire, so we returned fire, destroying the RPG gunners and the buildings in which they were located. Once all fires had ceased, the infantry began their push again clearing buildings, with tank support for security. We would continue this type of action for three more days until we reached the edge of the city on 16 November.

On 16 November, the infantry was ordered to conduct detailed clearing operations with tank support. My tank platoon continued to be assigned to 1/8 for the duration of

November and December conducting clearing operations along with patrols.

It has been said that this fight was the toughest fight since Khe Sanh. I personally cannot and will not make this comparison, however the determination that I will make regarding both fights is that they were both tough, bloody battles in which the lives of great Marines were lost. I can also say that I am proud of every Marine that I have ever served with and every Marine that has ever served. I am grateful every day that the Lord allowed me to come home to my family and that I have the opportunity to share this story with you. It is only because of God's grace and the fearless Marines which fought on my left and my right that I am here today. We are a nation feared by all because of our Marines.

Semper Fi and God Bless. Jeff Lee / Capt / USMC

Captain Lee

Gunnery Sergeant Chris Juhls

Platoon Sergeant of 2nd Tank Battalion, "C" Company, 2nd Platoon

Written by G/Sgt. Juhls

The trip over was long and I slept most of the way. We got off the plane in Main and again in Frankfort, Germany. They kept us separated from the civilians in both airports for good reason. We were all tense and it would probably cause problems. Our next stop was in Kuwait. The heat was intense as we waited to get on busses for Camp Victory.

The drive wasn't too bad as we had a military escort. Once we got there, it wasn't too long before we had our "welcome to War briefing" to let us know the "Rules of Engagement" (ROE). We spent a few days there in the sweltering heat, eating well and playing ping-pong at the Rec. Center.

Next it was on busses again to an airport where we loaded on a C-130 and took off. We loaded on trucks heading for Camp Fallujah, which seemed to take hours because IED'S were an issue the whole trip.

It was daylight by the time we finally arrived in our theater of operations. We stayed in Hajji in tents for four or five days until the trailers were cleared out for us. All the tank commanders (TC's) were assigned to the tank crews they were taking over. I was fortunate to replace SSGT Willenbecker. It was great to see him again and I knew I was in good hands.

Our first mission was to cover a bridge north west of Fallujah for a convoy carrying a large generator en route to Bagdad. The 3rd tank sheared an arm so we worked on that for the majority of the day. Once we were done, SSGt Will's gunner was standing on top of the turret. I will quote SSgt Will "Hey shit head, why don't you get down? You make a good target for a sniper." Seconds after his gunner crouched down- Bang! Sure as hell, we were getting rounds fired at us from a sniper. That was the first time in my life I was ever shot at, but it's amazing how fast you know it. SSgt. Will and the gunner jumped into the tank. Because I was on the back deck, I leaped off the back and onto the ground grabbing my Kevlar somehow in the process. A total of five rounds were fired, all of them too close for comfort. There is a certain cracking noise when bullets are close and all five were making it. (Sounded like the butts at the rifle range)

After the 5th round I crawled back on top of the tank in one fluid motion and dropped into the loaders hatch. We all got a pretty good laugh about it a little later. When all was quiet we set up a watch and passed out on top of the tank for the evening. I was not a virgin anymore but the worst was coming soon.

I did a couple of stints at the Traffic Control Point (TCP) just past the clover leaf. It was uneventful but later would become A Favorite Killing Grounds.

First Tanks finally left and it was our turn to take over. We continued to cover the TCP in a four day rotation with the other platoons in the company.

Our first rotation out there we got the call that a small observation point (OP) south of Fallujah, was taking mortar, RPG, and small arms fire. Here is where White Bravo Section became infamous and, no shit, they called us Cowboys. We headed down Route Mobile South to a white illumination pop up. Once there, a Marine guided us into position and showed us the house the fire was coming from. We set up, our thermals were on. It was about 2000. My gunner picked up men running outside the house in seconds after arriving. My left knee would not stop shaking. I had to use both hands to stop it. Cpl. Weir switched to coax and I told him to fire.

The white objects in the night started to dance from the incoming rounds and one was hit. I watched as his body went down and he slumped against the porch. Not to be outdone I grabbed the override and fired some myself. It was funny watching these objects run into each other as they tried to get into the house.

I told Cpl. Weir to go to the main gun. We loaded a heat round and waited for authorization to use it. It didn't take long and once it came I announced "fire". The first round entered by the front door. White 3 (SGT. Borel) also fired a heat and punctured the house. My second round entered to the right of the door. After just 3 main gun rounds, someone came over the radio and told us to cease fire.

We continued to scan with the thermals but there was nothing moving we just saw the guy propped up against the porch. A bunch of civilian cars drove up to the house minutes after we ceased fire and started dragging shit out of the house. My gunner swears it was body parts of some sort. We were all satisfied with our job and waited to be relieved to go back to the TCP. We were told the battalion Intel officer (S-2) needed to see us at once. It turns out that no one knew who gave the order for us to go there and who gave me the order to use the main gun. Once we arrived at the S-2, Sgt. Borel and I were grilled on the events that just took place. My commanding officer was there waiting on us.

After about an hour of the five W's, LtCol. Malay walked up to us and told us we did a good job and asked who we were. After the introductions I said with great pride "White Bravo". Col. Malay shook our hands and we left to relieve White Alpha at the TCP.

I said earlier that the TCP was a favorite killing field. There was a place called the soda factory south of the TCP that was ripe for the pickings. W-3 was estimated at killing at least 25 to 30 personnel in a month's time from there. My numbers were much smaller, probably 5 to 10.

One of these was when two guys at night were sitting on a dirt mound. Cpl Weir shot one and he rolled down the hill. The rest were main gun. So you can imagine trying to identify bodies after getting hit by a heat round.

Six Hour Fire Fight
Written by G/Sgt. Juhls

We just got to the TCP about 0800 that morning. 3/5 sent out a patrol north on Route Mobile the railroad trestle (OP Trestle). In a matter of minutes they were ambushed by a group of bad guys from the ghost town (eastern point Fallujah). I heard the traffic on the radio and volunteered our services.

They pulled us from the TCP (Alpha section relieved us) and moved us up on Mobile looking into the town. As soon as we got there, RPG's and small arms were dinging our tanks. It was incredible. We were engaged in our first major battle. Every weapons system on our tank was being used and they must have thought hell was here.

The enemy was courageous. They did not give up. Cpl Weir saw a guy in a window and literally cut him in half with the coax the top portion of his body fell out of the window like an old western. As we continued to lay waste, I heard a friendly voice come over the net. It was Gunny Hillard. He asked if he could join in the fun and I said, "By all means please do". He and SSgt. Higgins joined in the battle and it was on from there.

We continued to search and destroy the enemy. I remember looking back and seeing the infantry sitting down and drinking water and enjoying the show. I remember some Marine climbing on my tank and asking me to shoot a certain building. I told him to use the phone on the back. I later found out it was LtCol. Malay. Hillard got pulled to the north a little and engaged a cement factory to our rear. After he took that out, with the help of 500 lb bombs, he returned to us. I remember them dropping 500 pounders 50 meters to our front and artillery 75 meters to our front. I came across the radios and told them to shut it down. It was way too close and the tanks were rocking from the impacts.

It took six hours of continuous firing to get those bastards to back down. I don't know how many were killed but the ammo was free and they asked for it. We were the last element to leave and we were exhausted. After refitting back at Camp Baharia where we stayed during our TCP duty, we headed back out to the TCP so Alpha Section could get some rest. While pulling into the TCP and getting set "White 1" was hit by a mortar on his way out.

All I heard was, "I'm hit" and then Sgt. Alonzo (White 1 gunner) saying in a panic that the Lt. was hit.

I was thinking, "Was the tank hit or was the LT hit. Is he dead or what?" Once it was clarified that the Lt. was fucked up, SSgt. Hines took control of the situation and I radioed "Dark horse" all the info on the Lt. It turned out a mortar hit outside the TC's hatch, cut the LT's face pretty bad and injured the loader. Both were well deserving of the Purple Heart. Needless to say White Bravo got some revenge that night at the TCP killing quite a few of those bastards. The TCP went on like normal four day rotations per platoon 12 hour shifts per section. Night was better than day as there were more bad guys at night.

Sometime in early November we stopped the TCP and started all of our orders and mission briefs for the upcoming battle of Fallujah. We did, however, go North West up Mobile to help out recon. I went on a foot patrol with these guys. They were checking on routes into the city. It made me happy to know I was a tanker and I didn't have to walk everywhere.

During one of our order briefings, a 120 millimeter Chinese rocket soared over my head. I just got done taking a shit. The damned thing landed not 30 meters away and it was a dud. A lot of men would have been hurt including myself and Hilland if that thing blew up. Luck of the draw, I guess.

After many days at Camp Baharie and a lot of orders, I was told I was assigned to Kilo Company 3/5, 1st Marine Division (Samurai). I did not know it yet but we were about to make history.

The Push
Written by G/Sgt. Juhls

We got to our step off point on the 7th or 8th of Nov. It was west of Fallujah about two miles away. We slept on the tanks and were ready to move out the next morning. Our first objective was the Apartments. I was there once before and I took heavy fire from them on a scouting mission, with Lt. Col. Buhl and five other tanks. It was imperative that the Apartments get taken down for it will act as 3/5 Bn HQs and more importantly our aid station.

Morning came and it was time to move out. SGT. Borel led, I was right behind him and the rest of team Samurai followed. We hauled ass down the road (Route Golden) and came to the Apartments. We set up so we could cover the buildings and Route Golden coming into the city. Sgt. Borel shot his 50 cal. at a car that approached. A woman got out and started screaming, got back in and the car reentered Fallujah. The grunts stormed the Apartments and took them.

I got hit by one round of AK-47 but that was it. We were all relieved that what we thought would be a shootout turned out to be quite calm. We did not know or care at the time that White Bravo was the first tank to enter the city limits for the operation. Here is where the days are a blur.

W-3 broke down so we had to go to our combat trains. I think it was day two of the push. I sat in awe that night watching the city. There were explosions everywhere. The sounds were sometimes deafening and to make it better there was a thunder storm over the city. The bad guys had to have been thinking that Allah was pissed.

After getting Sgt Borel a new tank the next morning, we joined back up with Samurai in the city. Pictures of Hue City in Vietnam is the only way I can explain it, Tanks in front infantry clearing behind. That night I saw my first bastard. They caught him in a ditch. I was over helping Gunny Hillard (he was with L Co. 3/5 HAVOC) near a palm grove. The guy was all shot up in the legs so they put him on the front slope plate of Blue 3 and moved him back. The guy smelled like shit and was screaming from the pain.

I found myself getting off my tank to get a closer look. I had no remorse for this guy. If I were allowed, I would have shot him right there on the spot. We slept on the tanks that night. I had a meeting with Samurai CO (Capt. McNulty) that was about three blocks away. A sniper with the call sign "Banshee" came and got me and we walked through the alley ways of the city to the meeting. I felt fine knowing this guy was a sniper. What could happen?

That evening after the meeting we passed out on our tanks from total exhaustion. I was told by our radio watch that an RPG and some small arms impacted near the tanks and nobody flinched or woke up. We were tired.

The next couple of days were long and tiring. We patrolled down the center of the roads while the grunts went house to house. There were many fire fights. The enemy would watch a tank drive by and wait for the Marines to enter into the house. I think this was the point where I decided to break up my section and have the tanks go down separate lanes to support the grunts better. I know it was against doctrine but there was nothing in the manuals about good old Fallujah.

Each west/east running roads are named after females. Route Golden was April and we proceeded to clear south. By the 4th or 5th day we had made it to Donna. For some reason Sgt. Berel and I were stacked behind the other. I believe there were not enough lanes and Amtracks took up most of them. Once the days clearing was over the Samurai platoons were getting ready to go finish for the evening. The radio started going crazy and I knew the Marines got ambushed.

 Two alleyways to my west, or right, the Executive Officer (XO) called us over and we moved as fast as we could. There was a lot of confusion going on and I had no idea where they were. I saw vehicles to our south (Army) and some Hummers on Donna in my way. Finally, I found the alley only because I saw Marines kneeling down and a dead guy in the alley itself. I pulled in and Sgt. Borel followed. Sgt. Borel got hit by an RPG and a hand grenade. I didn't get hit by anything, but small arms. I told Sgt. Borel to get the hell out of the alley. I moved forward providing cover for some Marines who wanted to get better positions.

Once I was in the opening I tried to turn the tank around. I must have ripped down every wire and pole in the courtyard. By the time I was turned around the grunts handled their business and got the bastards. I was moved one more alley to the left when "Crazy Eye" (Lt. from Samurai) asked me to breech a hole in a wall. I leveled my gun and blasted a huge hole with a heat round. The day finally ended and we went back to resupply. After getting our supplies, ammo and fuel, we went back into town to try and find their fire base. We must have been out there for three hours trying to get to them.

At one point, I stopped and a large machine gun started lighting up the house directly in front of me. As the rounds were flying at us, I made the decision to get out of Dodge and bed down at the Apartments for the rest of the evening. The next day we pushed south again. I shot at three guys who came out of a house ten feet in front. It was their lucky day. I was so excited, I missed them by inches.

The grunts got them later when they couldn't breathe anymore. Cpl. Weir fired a heat round into the garage about 200 meters away and lit a car on fire. There were a lot of secondary's (explosions from ammo) which made us happy. Our tactics changed after the ambush a little. Now we were prepping the area of observation (AO) we did this by putting the main gun rounds into anything we thought was suspicious, which was everything. That afternoon I was told they needed two breeches made in a school courtyard to our south. I told the Lt. from Samurai that I wanted to do it while it was daylight.

My tank, by itself, rolled to the designated point and blasted two holes in the wall killing a guy in the process.

Sgt. Borel had an encounter with a retarded Iraqi earlier in the day. This guy just started running towards him and the grunts. No one fired on him. The retard ran past everyone until someone finally close lined him and slammed him on the deck. They tell us they found the execution rooms where that American was beheaded. There's also a rumor they found someone cemented into the wall.

The next day was a good one. We were put into a courtyard of a school and were told to destroy anything we wanted. After I fired 2 rounds, Sgt. Borel hit a house that had 55 gallon drums full of fuel. The flame came back to the tanks and scared the hell out of everyone, especially the machine gunners directly to our right.

Capt. McNulty came over the NET and asked "What the hell was that?"

The response was, "Tanks". I told Sgt. Borel to watch the spreading fire because it caught the courtyard we were in on fire, too. It was the beginning of our competition of who could cause more fires. Sgt Borel has the advantage on me now but there were still a couple weeks to go.

The Graveyard
Written by G/Sgt. Juhls

We pushed to a graveyard south of Donna. I had the center
sector. Sgt. Borel had the left sector. We started to prep the
buildings to our front which were on Elizabeth. I think
W-3 took out a machine gun team in his AO and it got a
great BDA (battle damage assessment). I was told that I
killed eight in a house and there were body parts all over,
but the best part was they found a heart in the center of the
room. Someone also killed a family except for the mother.
I don't know who did it, but they should not have been
there in the first place. By this point everyone was just
being ruthless. All we wanted to do was get this over with
and kill everything we saw. At the cemetery my driver got
a little action, he opened his hatch and fired his 9MM just
for fun.

The next couple of days were about the same, destroying
houses, grunts clearing and of course Sgt. Borel causing
major fires. We went to Heather which was the last west-
east running road Samurai had to take. Everyone let out a
sigh of relief. It was humbling in a way. Some Marines
had died and I'd walked over to the dead enemy. It was
nothing to see a cat or dog feeding on the guys in the
streets. The town was devastated, but it had to be done.

The relief did not last long though; we were told we had to
re-clear everything we just cleared this time pushing back
north. On the second day of the re-clearing and after a few
more Marines were killed or wounded, I was ordered

down south to a place named Queens to link up with Alpha White section. My days with Samurai were over, but not for long. The first day in Queens went smooth, we blew shit up, then grunts cleared. Five POW's surrendered to my front. They were scared shitless.

Day two was not much different from the first, however when we had to transfer ammo I got my finger stuck in the semi-redi ammo door and ripped it to shreds. The skin was over my nail and the bone was visible. The corpsman asked if I wanted Morphine but I declined saying I had to shit. On the way back to the train station I jumped off the tank and shit right in the street. I was laid up for ten days back at camp, losing my mind while the fighting was still going on.

From what I heard, my platoon did fine and it was almost over when I was well enough to join up with my section again. We were moved to Henry (all north and south roads were men's names); we set up a blocking position for a couple of days oriented south. Here is where I enjoyed my first Iraqi meal (chicken, rice and okra) along with some Chi tea. It beat the hell out of MRE's and the Iraqi soldiers made a lot of it. Once done with that, we headed back to our good buddies Samurai.

Back at Samurai
Written by G/Sgt. Juhls

I remember walking into their fire base north of Fran on Abe and getting greeted by smiles and handshakes. I went to see Crazy Eye. Sgt Borel and I walked into his room and the guy started crying. He gave us both hugs and said he was happy to see us again. It turns out he lost five Marines the day prior and was overcome with emotion.

We set up one block from the fire base to the north and the mission was to clean the houses that Marines were killed in. I sat in front of the first house for 2-3 hours before the grunts entered. Sure as shit, there were bad guys in there waiting on them. One Marine got shot in the arm and a corpsman broke a leg when a wall fell on him. I pulled back and unloaded at least 5-6 heat rounds into that house. For unknown reasons the enemy did not shoot at me or my loader the whole time we sat there.

If he would have we would have been dead. We were only a couple of meters away. We were told to back up into the fire base where we watched 500 lb bombs decimate the row of houses. After the bombs dropped, we all lined up and cleared the rest of the houses to Bill. That evening we set up a screen concentrated to the north of Bill. It had to be the most miserable night of this operation. It was cold, very cold, and you can't get comfortable in the TC's seat. At 0600 the next morning I pushed our tanks past the clover leaf where we all got out and drank coffee and warmed ourselves by way of the tank exhaust. The next couple of days were the same old, same old.

We pushed forward and the grunts re-cleared. Samurai took some more casualties but we were making progress. The grunts had a new tactic. They filled plastic water bottles with gas and soap with some Styrofoam mixed it up and placed a flash bang to it. They would find a room with bad guys and toss it in. It was very effective however someone got wind of it high up and made them stop. I personally thought it was a great idea but who am I. Hitting April was a great feeling. We were at the end of the line. All clearing had ceased and we were sent back to relax a little... or so we thought.

After Operations
Written by G/Sgt. Juhls

By now, it was December. We really had no rest, just a few sporadic days back at Camp Fallujah to pull maintenance and unwind. The new mission was to support the battalions on QRF (quick reaction force) and conduct patrols with them. We went on a rotation between 3/1, 1/3, and 3/5. I was fortunate enough to go to the castle. I think it was "K" Co. 3/1 (Spartan) these were the same guys I fought with when I fucked up my finger. It took me a while to figure out why they called themselves Spartans. It turns out that Capt. Gent the C.O. was from Sparta N. J.

We didn't do any patrols there, it was just a good place to get away from Camp Fallujah for a couple of days. Our other stop at rotation was right back at Samurai, there we patrolled a lot. Things in the city were quiet so we had the opportunity to see the Iraqi people trying to rebuild their lives. The kids would run up to the tanks and yell MISTA, MISTA, and point to their mouths and we would throw them candy or MRE's just to shut them up, but once you threw some kid something it was like kicking a fire ant hill, they all came. We spent more time smiling and waving on these patrols than anything else. Once in a while you would hear some Sparti gun fire, but it was usually the Iraqi soldiers shooting dogs or cats.

Election Day
Written by G/Sgt. Juhls

3/1 moved out and 3/4 from 29 Palms took over at the Castle. I don't remember the call sign. We were only with them a short time. During the elections they wanted to keep us stationary at the same spot. I told 3/4 operations chief that this was not a good idea. I said I would drive around showing a presence and occasionally go back to the Castle to check oils and such (drink coffee and play cards).

We only heard a mortar round that day and it was across the bridge west of the city. That same day I got the news about the helicopter crash and many of the 1/3 Marines were killed. I mention this because I was with them for a couple of days at the soda factory.

I understand all went well for the elections but from what I saw not a lot of Iraqi's voted. Things continued to be the same for a while. We were either sitting on our asses or patrolling. I did have a great Christmas dinner with 3/1 before they moved out.

The New Mission
Written by G/Sgt. Juhls

This place should be called bad karma. The first time out here, a Hummer hit an IED. The next day they went out to investigate it, and a seven ton was flipped, killing one Marine and injuring nine or ten more. I don't much care for this place. It's full of bad guys and the unit I'm with (3/8 out of Lejeune) is just not like the others.

Yesterday, three mortar rounds impacted in the fire base. One of them hit five meters from my tank. All is well though. We were inside a house, watching a movie and chilling out. There are a lot of things I failed to mention, like the seven I killed in a house off of Fran or how I had to run in front of unclear houses to get to a meeting. I have seen a lot of shit that no one should go through. The dead stick out the most. It was reassuring to see dead enemy, but when you sit down and think about the Marines who died, you wonder if there was something you could have done to prevent it. By no means will I beat myself up over this.

I guess the question is, has this changed my perspective on life or the way I will be in years to come? My answer is, I don't know; Only time will tell.

I do know that through all this, I will always have a beautiful woman waiting for me when I return along with two wonderful children.

Gunnery Sergeant Juhls retired from the Marine Corps on October 1, 2010

Left to Right, Ralph Schwartz, Jimmy Wendling, Rick Bayshore,
The Commandant of the Marine Corps General James F. Amos,
Clyde Hoch, Jan Wending

Left to Right, Jimmy and Brother Jan Wendling,
The Sergeant Major of the Marine Corps Michael Barrett,
Clyde Hoch, Todd Phillips

These photos were taken in Washington D. C. on November 10th 2012 at the Iwo Jima Monument. It was the 237th Marine Corps Birthday. I included the photos in the book because every year we get a little heavier and there are less of us. There were a total of 2,700,000 to have served in Vietnam. Supposedly, at the time this book was compiled, there were 450,000 Vietnam Veterans still living.

On November 10th Todd Phillips, Linda Holly (Jerry Holly's wife) and I have a toast to Jerry. Jerry served with Todd and I in Vietnam on the same tank. He succumbed to cancer (Agent Orange related), on the Marine Corps Birthday November 10th, 2006.

All of us in the photos except for the Commandant, Sgt Major and Rick Bashore were Marine Corps Vietnam Tankers and all of us saw our share of combat. Rick Bashore is an Army Veteran.

The USMC Tankers Association was established July 4, 1999. It is IRS sanctioned 501C vietnamnonprofit.org. Our members served in the US Marine Corps in tank and or Ontos units during the time of the war in the Republic of Vietnam between 1965 and 1970. The USMC Vietnam Tankers Association publishes a 48 page trimester magazine full of stories and photos of our time in-country. We conduct biannual reunions in various geographic locations all around our wonderful nation. We encourage and foster the retelling of our personal histories. Please visit our website at http//www.usmcvta.org.